HOLT®

ChemFile® A

Skills Practice Experiments

HOLT, RINEHART AND WINSTON

A Harcourt Education Company

Orlando • **Austin** • New York • San Diego • Toronto • London

Contents

Introduction to the Lab Program

Structure of the Experiments

INTRODUCTION

The opening paragraphs set the theme for the experiment and summarize its major concepts.

OBJECTIVES

Objectives highlight the key concepts to be learned in the experiment and emphasize the science process skills and techniques of scientific inquiry.

MATERIALS

These lists enable you to organize all apparatus and materials needed to perform the experiment. Knowing the concentrations of solutions is vital. You often need this information to perform calculations and to answer the questions at the end of the experiment.

SAFETY

Safety cautions are placed at the beginning of the experiment to alert you to procedures that may require special care. Before you begin, you should review the safety issues that apply to the experiment.

PROCEDURE

By following the procedures of an experiment, you perform concrete laboratory operations that duplicate the fact-gathering techniques used by professional chemists. You learn skills in the laboratory. The procedures tell you how and where to record observations and data.

DATA AND CALCULATIONS TABLES

The data that you collect during each experiment should be recorded in the labeled Data Tables provided. The entries you make in a Calculations Table emphasize the mathematical, physical, and chemical relationships that exist among the accumulated data. Both types of tables should help you to think logically and to formulate your conclusions about what occurs during the experiment.

CALCULATIONS

Space is provided for all computations based on the data you gather.

QUESTIONS

Based on the data and calculations, you should be able to develop plausible explanations for the phenomena you observe during the experiment. Specific questions require you to draw on the concepts you learn.

GENERAL CONCLUSIONS

This section asks broader questions that bring together the results and conclusions of the experiment and relate them to other situations.

Safety in the Chemistry Laboratory

CHEMICALS ARE NOT TOYS.

Any chemical can be dangerous if it is misused. Always follow the instructions for the experiment. Pay close attention to the safety notes. Do not do anything differently unless told to do so by your teacher.

Chemicals, even water, can cause harm. The trick is to know how to use chemicals correctly so that they will not cause harm. You can do this by following the rules on these pages, paying attention to your teacher's directions, and following the cautions on chemical labels and experiments.

These safety rules always apply in the lab.

1. **Always wear a lab apron and safety goggles.**

   Even if you aren't working on an experiment, laboratories contain chemicals that can damage your clothing, so wear your apron and keep the strings of the apron tied. Because chemicals can cause eye damage, even blindness, you must wear safety goggles. If your safety goggles are uncomfortable or get clouded up, ask your teacher for help. Try lengthening the strap a bit, washing the goggles with soap and warm water, or using an antifog spray.

2. **Generally, no contact lenses are allowed in the lab.**

   Even while wearing safety goggles, you can get chemicals between contact lenses and your eyes, and they can cause irreparable eye damage. If your doctor requires that you wear contact lenses instead of glasses, then you may need to wear special safety goggles in the lab. Ask your doctor or your teacher about them.

3. **Never work alone in the laboratory.**

   You should always do lab work under the supervision of your teacher.

4. **Wear the right clothing for lab work.**

   Necklaces, neckties, dangling jewelry, long hair, and loose clothing can cause you to knock things over or catch items on fire. Tuck in a necktie or take it off. Do not wear a necklace or other dangling jewelry, including hanging earrings. It isn't necessary, but it might be a good idea to remove your wristwatch so that it is not damaged by a chemical splash.

   Pull back long hair, and tie it in place. Nylon and polyester fabrics burn and melt more readily than cotton, so wear cotton clothing if you can. It's best to wear fitted garments, but if your clothing is loose or baggy, tuck it in or tie it back so that it does not get in the way or catch on fire.

   Wear shoes that will protect your feet from chemical spills—no open-toed shoes, sandals, or shoes made of woven leather straps. Shoes made of solid leather or a polymer are much better than shoes made of cloth. Also, wear pants, not shorts or skirts.

5. **Only books and notebooks needed for the experiment should be in the lab.**

   Do not bring other textbooks, purses, bookbags, backpacks, or other items into the lab; keep these things in your desk or locker.

6. **Read the entire experiment before entering the lab.**
Memorize the safety precautions. Be familiar with the instructions for the experiment. Only materials and equipment authorized by your teacher should be used. When you do the lab work, follow the instructions and the safety precautions described in the directions for the experiment.

7. **Read chemical labels.**
Follow the instructions and safety precautions stated on the labels. Know the location of Material Safety Data Sheets for chemicals.

8. **Walk carefully in the lab.**
Sometimes you will carry chemicals from the supply station to your lab station. Avoid bumping other students and spilling the chemicals. Stay at your lab station at other times.

9. **Food, beverages, chewing gum, cosmetics, and tobacco are *never* allowed in the lab.**
You already know this.

10. **Never taste chemicals or touch them with your bare hands.**
Also, keep your hands away from your face and mouth while working, even if you are wearing gloves.

11. **Use a sparker to light a Bunsen burner.**
Do not use matches. Be sure that all gas valves are turned off and that all hot plates are turned off and unplugged before you leave the lab.

12. **Be careful with hot plates, Bunsen burners, and other heat sources.**
Keep your body and clothing away from flames. Do not touch a hot plate just after it has been turned off. It is probably hotter than you think. Use tongs to heat glassware, crucibles, and other things and to remove them from a hot plate, a drying oven, or the flame of a Bunsen burner.

13. **Do not use electrical equipment with frayed or twisted cords or wires.**

14. **Be sure your hands are dry before you use electrical equipment.**
Before plugging an electrical cord into a socket, be sure the electrical equipment is turned off. When you are finished with it, turn it off. Before you leave the lab, unplug it, but be sure to turn it off first.

15. **Do not let electrical cords dangle from work stations; dangling cords can cause tripping or electric shocks.**
The area under and around electrical equipment should be dry; cords should not lie in puddles of spilled liquid.

16. **Know fire drill procedures and the locations of exits.**

17. **Know the locations and operation of safety showers and eyewash stations.**

18. **If your clothes catch on fire, *walk* to the safety shower, stand under it, and turn it on.**

19. **If you get a chemical in your eyes, walk immediately to the eyewash station, turn it on, and lower your head so that your eyes are in the running water.**
Hold your eyelids open with your thumbs and fingers, and roll your eyeballs around. You have to flush your eyes continuously for at least 15 min. Call your teacher while you are doing this.

20. **If you have a spill on the floor or lab bench, don't try to clean it up by yourself.**
First, ask your teacher if it is OK for you to do the cleanup; if it is not, your teacher will know how the spill should be cleaned up safely.

21. **If you spill a chemical on your skin, wash it off under the sink faucet, and call your teacher.**
If you spill a solid chemical on your clothing, brush it off carefully so that you do not scatter it, and call your teacher. If you get a liquid chemical on your clothing, wash it off right away if you can get it under the sink faucet, and call your teacher. If the spill is on clothing that will not fit under the sink faucet, use the safety shower. Remove the affected clothing while under the shower, and call your teacher. (It may be temporarily embarrassing to remove your clothing in front of your class, but failing to flush that chemical off your skin could cause permanent damage.)

22. **The best way to prevent an accident is to stop it before it happens.**
If you have a close call, tell your teacher so that you and your teacher can find a way to prevent it from happening again. Otherwise, the next time, it could be a harmful accident instead of just a close call.

23. **All accidents should be reported to your teacher, no matter how minor.**
Also, if you get a headache, feel sick to your stomach, or feel dizzy, tell your teacher immediately.

24. **For all chemicals, take only what you need.**
On the other hand, if you do happen to take too much and have some left over, **do not** put it back in the bottle. If somebody accidentally puts a chemical into the wrong bottle, the next person to use it will have a contaminated sample. Ask your teacher what to do with any leftover chemicals.

25. ***Never* take any chemicals out of the lab.**
You should already know this rule.

26. **Horseplay and fooling around in the lab are very dangerous.**
Never be a clown in the laboratory.

27. **Keep your work area clean and tidy.**
After your work is done, clean your work area and all equipment.

28. **Always wash your hands with soap and water before you leave the lab.**

29. **Whether or not the lab instructions remind you, *all* of these rules *apply all of the time*.**

QUIZ

Determine which safety rules apply to the following.

- Tie back long hair, and confine loose clothing. (Rule ? applies.)

- Never reach across an open flame. (Rule ? applies.)

- Use proper procedures when lighting Bunsen burners. Turn off hot plates and Bunsen burners that are not in use. (Rule ? applies.)

- Be familiar with the procedures and know the safety precautions before you begin. (Rule ? applies.)

- Use tongs when heating containers. Never hold or touch containers with your hands while heating them. Always allow heated materials to cool before handling them. (Rule ? applies.)

- Turn off gas valves that are not in use. (Rule ? applies.)

SAFETY SYMBOLS

To highlight specific types of precautions, the following symbols are used in the experiments. Remember that no matter what safety symbols and instructions appear in each experiment, all of the 29 safety rules described previously should be followed at all times.

EYE AND CLOTHING PROTECTION

- Wear safety goggles in the laboratory at all times. Know how to use the eyewash station.

- Wear laboratory aprons in the laboratory. Keep the apron strings tied so that they do not dangle.

CHEMICAL SAFETY

- Never taste, eat, or swallow any chemicals in the laboratory. Do not eat or drink any food from laboratory containers. Beakers are not cups, and evaporating dishes are not bowls.

- Never return unused chemicals to their original containers.

- Some chemicals are harmful to the environment. You can help protect the environment by following the instructions for proper disposal.

- It helps to label the beakers and test tubes containing chemicals.

- Never transfer substances by sucking on a pipet or straw; use a suction bulb.

- Never place glassware, containers of chemicals, or anything else near the edges of a lab bench or table.

CAUSTIC SUBSTANCES

- If a chemical gets on your skin or clothing or in your eyes, rinse it immediately, and alert your teacher.

- If a chemical is spilled on the floor or lab bench, tell your teacher, but do not clean it up yourself unless your teacher says it is OK to do so.

HEATING SAFETY

- When heating a chemical in a test tube, always point the open end of the test tube away from yourself and other people.

EXPLOSION PRECAUTION

- Use flammable liquids in small amounts only.
- When working with flammable liquids, be sure that no one else in the lab is using a lit Bunsen burner or plans to use one. Make sure there are no other heat sources present.

HAND SAFETY

- Always wear gloves or use cloths to protect your hands when cutting, fire polishing, or bending hot glass tubing. Keep cloths clear of any flames.
- Never force glass tubing into rubber tubing, rubber stoppers, or corks. To protect your hands, wear heavy leather gloves or wrap toweling around the glass and the tubing, stopper, or cork, and gently push the glass tubing into the rubber or cork.
- Use tongs when heating test tubes. Never hold a test tube in your hand to heat it.
- Always allow hot glassware to cool before you handle it.

GLASSWARE SAFETY

- Check the condition of glassware before and after using it. Inform your teacher of any broken, chipped, or cracked glassware because it should not be used.
- Do not pick up broken glass with your bare hands. Place broken glass in a specially designated disposal container.

GAS PRECAUTION

- Do not inhale fumes directly. When instructed to smell a substance, waft it toward you. That is, use your hand to wave the fumes toward your nose. Inhale gently.

RADIATION PRECAUTION

- Always wear gloves when handling a radioactive source.
- Always wear safety goggles when performing experiments with radioactive materials.
- Always wash your hands and arms thoroughly after working with radioactive materials.

HYGIENIC CARE

- Keep your hands away from your face and mouth.
- Always wash your hands before leaving the laboratory.

Any time you see any of the safety symbols, you should remember that all 29 of the numbered laboratory rules always apply.

Labeling of Chemicals

In any science laboratory the labeling of chemical containers, reagent bottles, and equipment is essential for safe operations. Proper labeling can lower the potential for accidents that occur as a result of misuse. Read labels and equipment instructions several times before you use chemicals or equipment. Be sure that you are using the correct items, that you know how to use them, and that you are aware of any hazards or precautions associated with their use.

All chemical containers and reagent bottles should be labeled prominently and accurately with labeling materials that are not affected by chemicals.

Chemical labels should contain the following information:

1. **Name of the chemical and its chemical formula**
2. **Statement of possible hazards** This is indicated by the use of an appropriate signal word, such as *DANGER*, *WARNING*, or *CAUTION*. This signal word usually is accompanied by a word that indicates the type of hazard present, such as *POISON*, *CAUSES BURNS*, *EXPLOSIVE*, or *FLAMMABLE*. Note that this labeling should not take the place of reading the appropriate Material Safety Data Sheet for a chemical.
3. **Precautionary measures** Precautionary measures describe how users can avoid injury from the hazards listed on the label. Examples include: "use only with adequate ventilation" and "do not get in eyes or on skin or clothing."
4. **Instructions in case of contact or exposure** If accidental contact or exposure does occur, immediate first-aid measures can minimize injury. For example, the label on a bottle of acid should include this instruction: "In case of contact, flush with large amounts of water; for eyes, rinse freely with water for 15 min and get medical attention immediately."
5. **The date of preparation and the name of the person who prepared the chemical** This information is important for maintaining a safe chemical inventory.

Suggested Labeling Scheme	
Name of contents	hydrochloric acid
Chemical formula and concentration or physical state	6 M HCl
Statements of possible hazards and precautionary and measures	WARNING! CAUSTIC and CORROSIVE—CAUSES BURNS Avoid contact with skin and eyes Avoid breathing vapors.
Hazard Instructions for contact or overexposure	IN CASE OF CONTACT: Immediately flush skin or eyes with large amounts of water for at least 15 min; for eyes, get medical attention immediately!
Date prepared or obtained Manufacturer (commercially obtained) or preparer (locally made)	May 8, 2005 Prepared by Betsy Byron, Faribault High School, Faribault, Minnesota

Laboratory Techniques

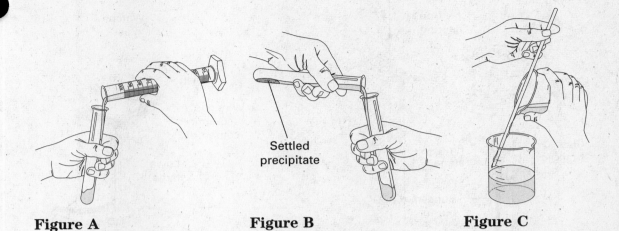

Figure A   **Figure B**   **Figure C**

Settled precipitate

DECANTING AND TRANSFERRING LIQUIDS

1. The safest way to transfer a liquid from a graduated cylinder to a test tube is shown in **Figure A**. Transfer the liquid at arm's length with your elbows slightly bent. This position enables you to see what you are doing and still maintain steady control.

2. Sometimes liquids contain particles of insoluble solids that sink to the bottom of a test tube or beaker. Use one of the methods given below to separate a supernatant (the clear fluid) from insoluble solids.

   a. **Figure B** shows the proper method of decanting a supernatant liquid in a test tube.

   b. **Figure C** shows the proper method of decanting a supernatant liquid in a beaker by using a stirring rod. The rod should touch the wall of the receiving container. Hold the stirring rod against the lip of the beaker containing the supernatant liquid. As you pour, the liquid will run down the rod and fall into the beaker resting below. Using this method will prevent the liquid from running down the side of the beaker you are pouring from.

HEATING SUBSTANCES AND EVAPORATING SOLUTIONS

1. Use care in selecting glassware for high-temperature heating. The glassware should be heat resistant.

2. When using a gas flame to heat glassware, use a ceramic-centered wire gauze to protect glassware from direct contact with the flame. Wire gauzes can withstand extremely high temperatures and will help prevent glassware from breaking. **Figure D** shows the proper setup for evaporating a solution over a water bath.

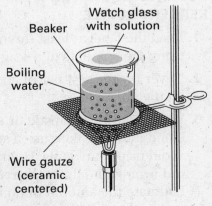

Watch glass with solution

Beaker

Boiling water

Wire gauze (ceramic centered)

Figure D

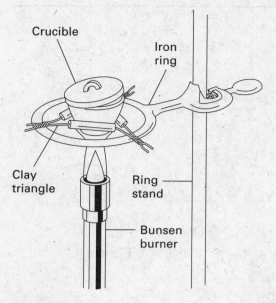

Figure E

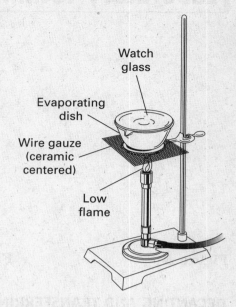

Figure F

3. In some experiments you are required to heat a substance to high temperatures in a porcelain crucible. **Figure E** shows the proper apparatus setup used to accomplish this task.

4. **Figure F** shows the proper setup for evaporating a solution in a porcelain evaporating dish with a watch glass cover that prevents spattering.

5. Glassware, porcelain, and iron rings that have been heated may look cool after they are removed from a heat source, but they can burn your skin even after several minutes of cooling. Use tongs, test tube holders, or heat-resistant mitts and pads whenever you handle this apparatus.

6. You can test the temperature of questionable beakers, ring stands, wire gauzes, or other pieces of apparatus that have been heated, by holding the back of your hand close to their surfaces before grasping them. You will be able to feel any heat generated from the hot surfaces. **Do not touch the apparatus until it is cool.**

POURING LIQUID FROM A REAGENT BOTTLE

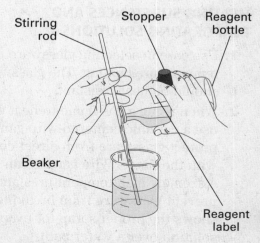

1. Read the label at least three times before using the contents of a reagent bottle.

2. Never lay the stopper of a reagent bottle on the lab table.

3. When pouring a caustic or corrosive liquid into a beaker, use a stirring rod to avoid drips and spills. Hold the stirring rod against the lip of the reagent bottle. Estimate the amount of liquid you need, and pour this amount along the rod into the beaker. See **Figure G.**

Figure G

4. Take extra precautions when handling a bottle of acid or strong base. Remember the following important rules: Never add water to any concentrated acid, particularly sulfuric acid, because the mixture can splash and will generate a lot of heat. To dilute any acid, add the acid to water in small quantities, while stirring slowly. Remember the "triple A's"—Always Add Acid to water.

5. Examine the outside of the reagent bottle for any liquid that has dripped down the bottle or spilled on the counter top. Your teacher will show you the proper procedures for cleaning up a chemical spill.

6. Never pour reagents back into stock bottles. At the end of the experiment, your teacher will tell you how to dispose of any excess chemicals.

HEATING MATERIAL IN A TEST TUBE

1. Check to see that the test tube is heat resistant.

2. Always use a test-tube holder or clamp when heating a test tube.

3. Never point a heated test tube at anyone, because the liquid may splash out of the test tube.

4. Never look down into the test tube while heating it.

5. Heat the test tube from the upper portions of the tube downward and continuously move the test tube, as indicated in **Figure H.** Do not heat any one spot on the test tube. Otherwise, a pressure buildup may cause the bottom of the tube to blow out.

USING A MORTAR AND PESTLE

1. A mortar and pestle should be used for grinding only one substance at a time. See **Figure I.**

2. Never use a mortar and pestle for simultaneously mixing different substances.

3. Place the substance to be broken up into the mortar.

4. Firmly push on the pestle to crush the substance. Then grind it to pulverize it.

5. Remove the powdered substance with a porcelain spoon.

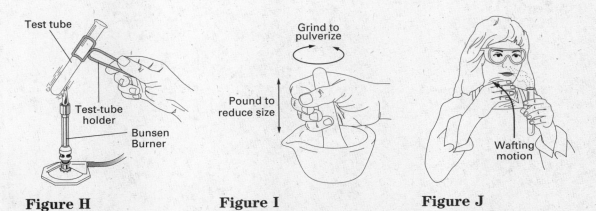

Figure H   Figure I   Figure J

DETECTING ODORS SAFELY

1. Test for the odor of gases by wafting your hand over the test tube and cautiously sniffing the fumes, as indicated in **Figure J.**

2. Do not inhale any fumes directly.

3. Use a fume hood whenever poisonous or irritating fumes are involved. **Do not** waft and sniff poisonous or irritating fumes.

Skills Practice

Laboratory Procedures

The best way to become familiar with chemical apparatus is to handle the pieces yourself in the laboratory. This experiment is divided into several parts in which you will learn how to adjust the gas burner, insert glass tubing into a rubber stopper, use a balance, handle solids, measure liquids, and filter a mixture. Great emphasis is placed on safety precautions that should be observed whenever you perform an experiment and use the apparatus. In many of the later experiments, references will be made to these "Laboratory Techniques." In later experiments you will also be referred to the safety precautions and procedures explained in all parts of this experiment. It is important that you develop a positive approach to a safe and healthful environment in the lab.

OBJECTIVES

Observe proper safety techniques with all laboratory equipment.

Use laboratory apparatus skillfully and efficiently.

Recognize the names and functions of all apparatus in the laboratory.

Develop a positive approach toward laboratory safety.

Always wear safety goggles, gloves, and a lab apron to protect your eyes and clothing. If you get a chemical in your eyes, immediately flush the chemical out at the eyewash station while calling to your teacher. Know the location of the emergency lab shower and eyewash station and the procedures for using them.

Do not touch any chemicals. If you get a chemical on your skin or clothing, wash the chemical off at the sink while calling to your teacher. Make sure you carefully read the labels and follow the precautions on all containers of chemicals that you use. If there are no precautions stated on the label, ask your teacher what precautions to follow. Do not taste any chemicals or items used in the laboratory. Never return leftover chemicals to their original containers; take only small amounts to avoid wasting supplies.

Do not heat glassware that is broken, chipped, or cracked. Use tongs or a hot mitt to handle heated glassware and other equipment because hot glassware does not always look hot.

When using a Bunsen burner, confine long hair and loose clothing. If your clothing catches on fire, WALK to the emergency lab shower and use it to put out the fire.

When heating a substance in a test tube, the mouth of the test tube should point away from where you and others are standing. Watch the test tube at all times to prevent the contents from boiling over.

| Laboratory Procedures *continued*

 Never put broken glass in a regular waste container. Broken glass should be disposed of separately according to your teacher's instructions.

When you insert glass tubing into stoppers, lubricate the glass with water or glycerin and protect your hands and fingers. Wear leather gloves or place folded cloth pads between both your hands and the glass tubing. Then *gently* push the tubing into the stopper hole. In the same way, protect your hands and fingers when removing glass tubing from stoppers and from rubber or plastic tubing.

PART 1—THE BURNER

MATERIALS

- Bunsen burner and related equipment
- copper wire, 18 gauge
- evaporating dish
- forceps
- heat-resistant gloves

- heat-resistant mat
- lab apron
- safety goggles
- sparker

Procedure

1. Put on safety goggles, and a lab apron. Use heat-resistant gloves when handling hot items.

2. The Bunsen burner is commonly used as a source of heat in the laboratory. Look at **Figure 1** as you examine your Bunsen burner and identify the parts. Although the details of construction vary among burners, each has a gas inlet located in the base, a vertical tube or barrel in which the gas is mixed with air, and adjustable openings or ports in the base of the barrel. These ports admit air to the gas stream. The burner may have an adjustable needle valve to regulate the flow of gas. In some models the gas flow is regulated simply by adjusting the gas valve on the supply line. The burner is always turned off at the gas valve, never at the needle valve.

   CAUTION: Before you light the burner, check to see that you and your partner have taken the following safety precautions against fires: Wear safety goggles, aprons, and heat-resistant gloves. Confine long hair and loose clothing: tie long hair at the back of the head and away from the front of the face, and roll up long sleeves on shirts, blouses, and sweaters away from the wrists. You should also know the locations of fire extinguishers, fire blankets, safety showers, and sand buckets and the procedure for using them in case of a fire.

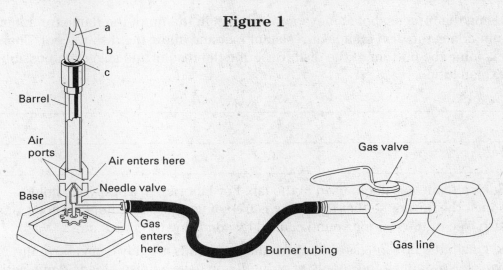

Figure 1

3. When lighting the burner, partially close the ports at the base of the barrel, turn the gas full on, hold the sparker about 5 cm above the top of the burner, and proceed to light. The gas flow may then be regulated by adjusting the gas valve until the flame has the desired height. If a very low flame is needed, remember that the ports should be partially closed when the gas pressure is reduced. Otherwise the flame may burn inside the base of the barrel. When the flame is improperly burning in this way, the barrel will get very hot, and the flame will produce a poisonous gas, carbon monoxide.

CAUTION: If the flame is burning inside the base of the barrel, immediately turn off the gas at the gas valve. Do not touch the barrel, because it is extremely hot. Allow the barrel of the burner to cool, and then proceed as follows:

Begin again, but first decrease the amount of air admitted to the burner by partially closing the ports. Turn the gas full on, and then relight the burner. Control the height of the flame by adjusting the gas valve. By taking these steps, you should acquire a flame that is burning safely and is easily regulated.

4. Once you have a flame that is burning safely and steadily, you can experiment by completely closing the ports at the base of the burner. What effect does this have on the flame?

_____

_____

_____

Laboratory Procedures *continued*

Using the forceps, hold an evaporating dish in the tip of the flame for about 3 min. Place the dish on a heat-resistant mat and allow the dish to cool. Then examine the bottom of the dish. Describe the results and suggest a possible explanation.

_____

_____

_____

Such a flame is seldom used in the lab. For laboratory work, you should adjust the burner so that the flame is free of yellow color, nonluminous, and also free of the roaring sound caused by admitting too much air.

5. Regulate the flow of gas so that the flame extends roughly 8 cm above the barrel. Now adjust the supply of air until you have a noisy, steady flame with a sharply defined, light-blue inner cone. This adjustment gives the highest temperature possible with your burner. Using the forceps, insert a 10-cm piece of copper wire into the flame just above the barrel. Lift the wire slowly up through parts of the flame. Where is the hottest portion of the flame located?

_____

_____

Hold the wire in this part of the flame for a few seconds. What happens?

_____

6. Shut off the gas burner. Now think about what you have just observed in steps **4** and **5.** Why is the nonluminous flame preferred over the yellow luminous flame in the laboratory?

_____

7. Clean the evaporating dish and put away the burner. All the equipment you store in the lab locker or drawer should be completely cool, clean, and dry. Be sure that the valve on the gas jet is completely shut off. Remember to wash your hands thoroughly with soap at the end of each laboratory period.

PART 2—GLASS MANIPULATION

MATERIALS

- cloth pads or leather gloves
- glass funnel
- lab apron
- rubber hose

- rubber stopper, 1-hole
- safety goggles
- water or glycerin

Procedure

1. Inserting glass tubing into rubber stoppers can be very dangerous. The following precautions should be observed to prevent injuries:

   a. Never attempt to insert glass tubing that has a jagged end. Glass tubing should be fire polished before it is inserted into a rubber stopper. To fire polish glass tubing, heat the end in a flame until the end is smooth. **Never fire polish anything without your teacher's permission, and proper supervision. Use tongs or a hot mitt to handle heated glassware and other equipment, because heated glassware does not look hot.**

   b. Use water or glycerin as a lubricant on the end of the glass tubing before inserting it into a rubber stopper. Ask your teacher for the proper lubricant. **CAUTION: Protect your hands and fingers when inserting glass tubing into a rubber stopper.**

   c. Wear leather gloves or place folded cloth pads between your hands and the glass tubing. Hold the glass tubing as close as possible to the part where it is to enter the rubber stopper. Always point the glass tubing away from the palm of your hand that holds the stopper, as shown in **Figure 2** below. Using a twisting motion, gently push the tubing into the stopper hole.

   Figure 2

   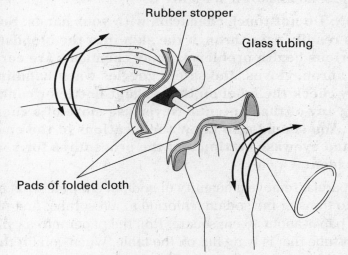

   Rubber stopper

   Glass tubing

   Pads of folded cloth

   d. At the end of the experiment, put on leather gloves, or place folded cloth pads between your hands and the glass tubing, and remove the rubber stoppers from the tubing to keep them from sticking or "freezing" to the glass. Use a lubricant, as directed in **step 1b,** if the stopper or tubing won't budge.

2. When inserting glass tubing into a rubber or plastic hose, observe the same precautions discussed in **steps 1a–1c.** The glass tubing should be lubricated before inserting it into the rubber or plastic hose. The rubber hose should be cut at an angle before the insertion of the glass tubing. The angled cut in the hose allows the rubber to stretch more readily.

   CAUTION: Protect your hands when inserting or removing glass tubing.

Laboratory Procedures *continued*

At the end of an experiment, immediately remove the glass tubing from the hose. When disassembling, follow the precautions that were given in **step 1d.**

Carefully follow these precautions and techniques whenever an experiment requires that you insert glass tubing into either a rubber stopper or a rubber or plastic hose. You will be referred to these safety precautions, wherever appropriate, throughout the lab course.

PART 3—HANDLING SOLIDS

MATERIALS

- glazed paper
- gloves
- lab apron
- safety goggles

- sodium chloride
- spatula
- test tube

Procedure

1. Solids are usually kept in wide-mouthed bottles. A spatula should be used to dip out the solid as shown in **Figure 3.**

   CAUTION: Do not touch chemicals with your hands. Some chemical reagents readily pass through the skin into the bloodstream and can cause serious health problems. Some chemicals are corrosive. Always wear an apron, gloves, and safety goggles when handling chemicals. Carefully check the label on the reagent bottle or container before removing any of the contents. Never use more of a chemical than directed. You should also know the locations of the emergency lab shower and eyewash station and the procedures for using them in case of an accident.

   Using a spatula, remove a quantity of sodium chloride from its reagent bottle. In order to transfer the sodium chloride to a test tube, first place it on a piece of glazed paper about 10 cm square. Roll the paper into a cylinder and slide it into a test tube that is lying flat on the table. When you lift the tube to a vertical position and tap the paper gently, the solid will slide down into the test tube, as shown in **Figure 4.**

   CAUTION: Never try to pour a solid from a bottle into a test tube. As a precaution against contamination, never pour unused chemicals back into their reagent bottles.

2. Dispose of the solid sodium chloride and the glazed paper in the waste jars or containers provided by your teacher.

   CAUTION: Never discard chemicals or broken glassware in the waste paper basket. This is an important safety precaution against fires, and it prevents personal injuries (such as hand cuts) to anyone who empties the wastepaper basket.

Laboratory Procedures *continued*

3. Remember to clean up the lab station and to wash your hands at the end of this part of the experiment.

Figure 3   **Figure 4**

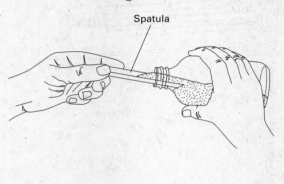
Spatula

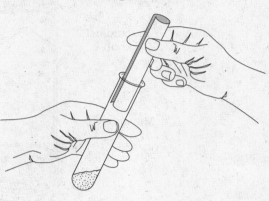

PART 4—THE BALANCE

MATERIALS

- balance, centigram
- glazed paper
- gloves
- lab apron

- safety goggles
- sodium chloride
- spatula
- weighing paper

Procedure

1. When a balance is required for determining mass, you will use a centigram balance like the one shown in **Figure 5.** The centigram balance has a readability of 0.01 g. This means that your mass readings should all be recorded to the nearest 0.01 g.

2. Before using the balance, always check to see if the pointer is resting at zero. If the pointer is not at zero, check the riders on the scales. If all the scale riders are at zero, turn the zero-adjust knob until the pointer rests at zero. The zero-adjust knob is usually located at the far left end of the balance beam, as shown in **Figure 5.** Note: The balance will not adjust to zero if the movable pan has been removed. **Never place chemicals or hot objects directly on the balance pan.** Always use weighing paper or a glass container. Chemicals can permanently damage the surface of the balance pan and affect the accuracy of measurements.

Laboratory Procedures *continued*

Figure 5

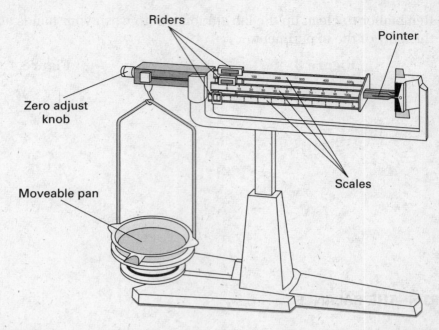

Riders

Pointer

Zero adjust knob

Scales

Moveable pan

3. In many experiments, you will be asked to determine the mass of a specified amount of a chemical solid. Use the following procedure to obtain approximately 13 g of sodium chloride.

a. Make sure that the pointer on the balance is set at zero. Obtain a piece of weighing paper and place it on the balance pan. Determine the mass of the paper by adjusting the riders on the various scales. Record the mass of the weighing paper to the nearest 0.01 g.

Mass of paper: _____

b. Add 13 grams to the balance by sliding the rider on the 100 g scale to 10 and the rider on the 10 g scale to 3.

c. Using a spatula, obtain a quantity of sodium chloride from the reagent bottle and place it on a separate piece of glazed paper.

d. Now slowly pour the sodium chloride from the glazed paper onto the weighing paper on the balance pan until the pointer once again comes to zero. In most cases, you will only have to be close to the specified mass. Do not waste time trying to obtain exactly 13.00 g. Instead, read the exact mass when the pointer rests close to zero and you have around 13 g of sodium chloride in the balance pan. The mass might be 13.18 g. Record your exact mass of sodium chloride and the weighing paper to the nearest 0.01 g. (Hint: Remember to subtract the mass of the weighing paper to find the mass of sodium chloride.)

Mass of NaCl and paper: _____

4. Wash your hands thoroughly with soap and water at the end of each lab period.

Laboratory Procedures *continued*

PART 5—MEASURING LIQUIDS

MATERIALS

- beaker, 50 mL
- beaker, 250 mL
- buret
- buret clamp
- gloves
- graduated cylinder, 100 mL

- lab apron
- pipet
- ring stand
- safety goggles
- water

Procedure

1. For approximate measurements of liquids, a graduated cylinder, such as the one shown in **Figure 6,** is generally used. These cylinders are usually graduated in milliliters (mL), reading from the bottom up. They may also have a second column of graduations reading from top to bottom. Examine your cylinder for these markings. Record the capacity and describe the scale of your cylinder in the space below.

   Observations: _____

   _____

Figure 6

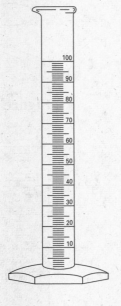

Figure 7

2. A pipet or a buret is used for more accurate volume measurements. Pipets, which are made in many sizes, are used to deliver measured volumes of liquids. A pipet is fitted with a suction bulb, as shown in **Figure 7.** The bulb is used to withdraw air from the pipet while drawing up the liquid to be measured.

   CAUTION: Always use the suction bulb. NEVER pipet by mouth.

3. Burets are used for delivering any desired
quantity of liquid up to the capacity of the buret.
Many burets are graduated in tenths of milliliters. When using a buret, follow these steps:

Figure 8

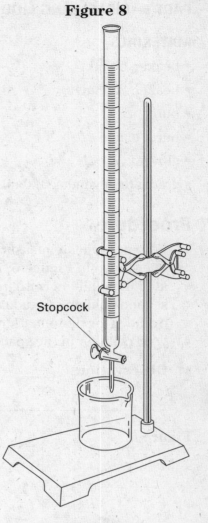

Stopcock

a. Clamp the buret in position on a ring stand, as
shown in **Figure 8.**

b. Place a 250-mL beaker under the tip of the
buret. The beaker serves to catch any liquid
that is released.

c. Pour a quantity of liquid that you want to
measure from the liquid's reagent bottle into a
50-mL beaker. (NOTE: In this first trial, you
will use water.) Pour the liquid from the
beaker into the top of the buret, being careful
to avoid spills. (NOTE: Always carefully
check the label of any reagent bottle before
removing any liquid).

**CAUTION: Never pour a liquid directly
from its reagent bottle into the buret.
You should first pour the liquid into a
small, clean, dry beaker (50 mL) that is
easy to handle. Then pour the liquid from
the small beaker into the buret. This
simple method will prevent unnecessary
spillage. Never pour any unused liquid
back into the reagent bottle.**

d. Fill the buret with the liquid and then open
the stopcock to release enough liquid to fill
the tip below the stopcock and bring the level
of the liquid within the scale. The height at
which the liquid stands is then read accurately. Practice this procedure several times
by pouring water into the buret and emptying
it through the stopcock.

Figure 9

26

A - - - - - - - - - - - - - D
C - - - - - - - - - - - - - B
27

28

4. Notice that the surface of a liquid in the buret is
slightly curved. It is concave if it wets the glass
and convex if it does not wet the glass. Such a
curved surface is called a meniscus. If a liquid
wets the glass, read the bottom of the meniscus,
as shown in **Figure 9.** This is the line *CB*. If you
read the markings at the top of the meniscus, *AD*, you will get an incorrect
reading. Locate the bottom of the meniscus and read the water level in your buret.

Buret reading: _____

Laboratory Procedures *continued*

5. After you have taken your first buret reading, open the stopcock to release some of the liquid. Then read the buret again. The exact amount released is equal to the difference between your first and final buret reading. Practice measuring liquids by measuring 10 mL of water, using a graduated cylinder, a pipet, and a buret.

6. At the end of this part of the experiment, the equipment you store in the lab locker or drawer should be clean, dry, and arranged in an orderly fashion for the next lab experiment.

CAUTION: In many experiments, you will have to dispose of a liquid chemical at the end of a lab. Always ask your teacher about the correct method of disposal. In many instances, liquid chemicals can be washed down the sink's drain by diluting them with plenty of tap water. Toxic chemicals should be handled only by your teacher. All apparatus should be washed, rinsed, and dried.

7. Remember to wash your hands thoroughly with soap at the end of this part of the experiment.

PART 6—FILTRATION

MATERIALS

- beaker, 250 mL (2)
- Bunsen burner and related equipment
- evaporating dish
- filter paper
- fine sand
- funnel
- glass stirring rod
- gloves
- iron ring
- lab apron
- ring stand
- safety goggles
- sodium chloride
- sparker
- wash bottle
- water
- wire gauze, ceramic-centered

Laboratory Procedures *continued*

Procedure

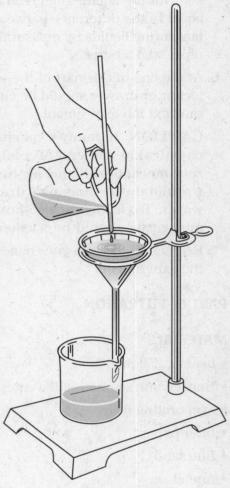

Figure 10

1. Sometimes liquids contain particles of insoluble solids that are present either as impurities or as precipitates formed by the interaction of the chemicals used in the experiment. If the particles are denser than water, they soon sink to the bottom. Most of the clear, supernatant liquid above the solid may be poured off without disturbing the precipitate. This method of separation is known as decantation.

2. Fine particles, or particles that settle slowly, are often separated from a liquid by filtration. Support a funnel on a small ring on the ring stand, as shown in **Figure 10.** Use a beaker to collect the filtrate. Adjust the funnel so that the stem of the funnel just touches the inside wall of the beaker.

3. Fold a circular piece of filter paper along its diameter, then fold it again to form a quadrant, as shown in **Figure 11.** Separate the folds of the filter paper, with three thickness on one side and one on the other; then place the resulting filter paper cone in the funnel.

Figure 11

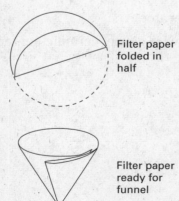

Filter paper folded in half

Filter paper folded in quarters

Filter paper ready for funnel

Filter paper in funnel

Laboratory Procedures *continued*

The funnel should be wet before you insert the filter paper. Use your plastic wash bottle to wet the funnel and the filter paper. Press the edges of the filter paper firmly against the sides of the funnel so no air can get between the funnel and the filter paper while the liquid is being filtered. *EXCEPTION: A filter should not be wetted with water when the liquid to be filtered does not mix with water. Why?*

_____

_____

4. Dissolve 2 or 3 g of sodium chloride in a beaker containing about 50 mL of water, and then stir into the solution an equal volume of fine sand. Filter out the sand by pouring the mixture into the filter, observing the following suggestions:

  a. The filter paper should not extend above the edge of the funnel. Use filter paper that leaves about 1 cm of the funnel exposed.

  b. Do not completely fill the funnel. It must never overflow.

  c. Try to establish a water column in the stem of the funnel to eliminate air bubbles, and then add the liquid quickly enough to keep the mixture level about 1 cm from the top of the filter paper.

  d. When a liquid is poured from a beaker, it may adhere to the glass and run down the outside wall. This may be avoided by holding a stirring rod against the lip of the beaker, as shown in **Figure 10** on the previous page. The liquid will run down the rod and drop off into the funnel without running down the outside of the beaker. The sand is retained on the filter paper. What property of the sand enables it to be separated from the water by filtration?

_____

What does the filtrate contain?

_____

5. The salt can be recovered from the filtrate by pouring the filtrate into an evaporating dish and evaporating it over a low flame nearly to dryness. **Figure 12** shows a correct setup for evaporation.

CAUTION: When using a Bunsen burner, confine loose clothing and long hair. Wear your safety goggles, lab apron, and heat-resistant gloves.

Laboratory Procedures *continued*

6. Remove the flame as soon as the liquid begins to spatter. Shut off the burner. What property of salt prevents it from being separated from the water by filtration?

_____

7. All equipment should be clean, dry, and put away in an orderly fashion for the next lab experiment. Be sure that the valve on the gas jet is completely shut off. Make certain that the filter papers and sand are disposed of in the waste jars or containers and not down the sink. Remember to wash your hands thoroughly with soap at the end of each laboratory period.

Figure 12

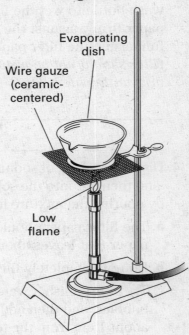

Evaporating dish

Wire gauze (ceramic-centered)

Low flame

Analysis

Answer the following questions in complete sentences.

1. Organizing Ideas As soon as you enter the lab, what safety equipment should you put on immediately?

_____

2. Organizing Ideas Before doing an experiment, what should you read and discuss with your teacher?

_____

_____

3. Organizing Ideas Before you light a burner, what safety precautions should always be followed?

_____

_____

_____

4. Organizing Ideas What immediate action should you take when the flame of your burner is burning inside the base of the barrel?

_____

_____

Laboratory Procedures *continued*

5. Organizing Ideas What type of flame is preferred for laboratory work, and why?

_____

_____

6. Analyzing Ideas When inserting glass tubing, why is it important that you wear safety goggles and gloves and that you cover the tubing and stopper with protective pads of cloth?

_____

_____

_____

7. Analyzing Ideas What do you think might be a common cause of fires in lab drawers or lockers?

_____

8. Analyzing Ideas Why are broken glassware, chemicals, matches, and other laboratory debris never discarded in a wastepaper basket?

_____

_____

_____

9. Organizing Ideas List the safety precautions that should be observed when inserting or removing glass tubing from a rubber stopper or rubber hose.

_____

_____

_____

_____

_____

_____

_____

_____

Laboratory Procedures *continued*

10. Analyzing Ideas Why should you never touch chemicals with your hands?

_____

_____

11. Organizing Ideas What precaution can help prevent chemical contamination in reagent bottles?

_____

12. Analyzing Ideas Why are chemicals and hot objects never placed directly on the balance pan?

_____

_____

_____

13. Organizing Ideas List three pieces of equipment used in the laboratory for measuring small quantities of liquids. What is the correct procedure for filling a buret with liquid?

_____

_____

_____

14. Organizing Ideas What is the rule about the size of filter paper to be used with a funnel?

_____

_____

_____

15. Organizing Ideas How can a liquid be transferred from a beaker to a funnel without spattering and without running down the outside wall of the beaker?

_____

_____

_____

_____

Laboratory Procedures *continued*

16. Organizing Ideas Describe the condition of all lab equipment at the end of an experiment. What should be checked before you leave the lab?

_____

_____

_____

17. Organizing Ideas What is the correct procedure for removing a solid reagent from its container in preparation for its use in an experiment?

_____

_____

_____

_____

_____

_____

18. Organizing Ideas What is the correct procedure for removing a liquid reagent from its container in preparation for its use in an experiment?

_____

_____

_____

_____

_____

_____

19. Analyzing Ideas Why is it important to use low flame when evaporating water from a recovered filtrate?

_____

_____

_____

| Laboratory Procedures *continued*

General Conclusions

SAFETY CHECK

Identify the following safety symbols:

a. _____

b. _____

c. _____

d. _____

e. _____

f. _____

g. _____

h. _____

i. _____

j. _____

LABELING

Practice labeling a chemical container or bottle by filling in the appropriate information that is missing from the label pictured on the following page. Use 6 M sodium hydroxide (NaOH) as the solution to be labeled. (Hint: 6 M sodium hydroxide is a caustic and corrosive solution, and it can be considered potentially as hazardous as 6 M HCl.)

SAMPLE LABEL

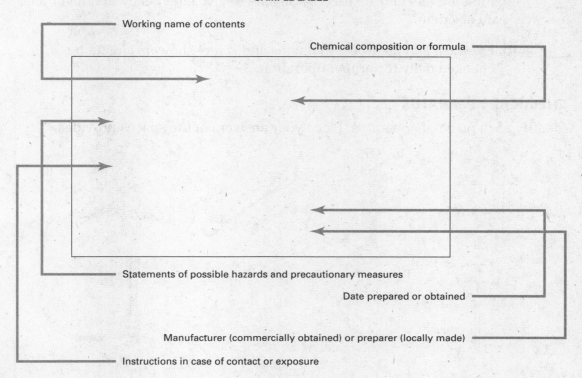

Working name of contents

Chemical composition or formula

Statements of possible hazards and precautionary measures

Date prepared or obtained

Manufacturer (commercially obtained) or preparer (locally made)

Instructions in case of contact or exposure

TRUE OR FALSE

Read the following statements and indicate whether they are true or false. Place your answer in the space next to the statement.

_____ **1.** Never work alone in the laboratory.

_____ **2.** Never lay the stopper of a reagent bottle on the lab table.

_____ **3.** At the end of an experiment, in order to save the school's money, save all excess chemicals and pour them back into their stock bottles.

_____ **4.** The quickest and safest way to heat a material in a test tube is by concentrating the flame on the bottom of the test tube.

_____ **5.** Use care in selecting glassware for high-temperature heating. Glassware should be Pyrex or a similar heat-treated type.

_____ **6.** A mortar and pestle should be used for grinding only one substance at a time.

_____ **7.** Safety goggles protect your eyes from particles and chemical injuries. It is completely safe to wear contact lenses under them while performing experiments.

_____ **8.** Never use the wastepaper basket for disposal of chemicals.

Name _____ Class _____ Date _____

Laboratory Procedures continued

_____ **9.** First aid kits may be used by anyone to give emergency treatment after an accident.

_____ **10.** Eyewash and facewash fountains and safety showers should be checked daily for proper operation.

CHEMICAL APPARATUS

Identify each piece of apparatus. Place your answers in the spaces provided.

a.

b.

c.

d.

e.

f.

g.

h.

i.

j.

k.

l.

a. _____   g. _____

b. _____   h. _____

c. _____   i. _____

d. _____   j. _____

e. _____   k. _____

f. _____   l. _____

Name _____ Class _____ Date _____

Accuracy and Precision in Measurements

In this experiment, you will determine the volume of a liquid in two different ways and compare the results. You will also calculate the density of a metal, using your measurements of its mass and volume.

$$D = \frac{m}{V}$$

You will compare your result with the accepted value found in a the handbook. The experimental error and percentage error in each part of the experiment will be calculated.

The *experimental error* is calculated by subtracting the accepted value from the observed or experimental value, as follows:

$$Experimental\ error = Experimental\ value - Accepted\ value$$

The *percentage error* is calculated according to the following equation:

$$Percentage\ error = \frac{Experimental\ value - Accepted\ value}{Accepted\ value} \times 100$$

The sign of the experimental error and the percentage error may be either positive (the experimental result is too high) or negative (the experimental result is too low).

You will determine the average value for the density of a metal by averaging the values obtained by the entire class. Using this value, you will calculate the experimental error and percentage error in the class average.

OBJECTIVES

Use experimental measurements in calculations.

Organize data by compiling it in tables.

Compute an average value from class data.

Recognize the importance of accuracy and precision in scientific measurements.

Relate the reliability of experimental data to uncertainty and percent error.

MATERIALS

- 15 cm plastic ruler
- 25 mL graduated cylinder
- 100 mL beaker
- 100 mL graduated cylinder
- balance
- metal shot (aluminum, copper, lead)
- thermometer, nonmercury, 0–100°C

| Accuracy and Precision in Measurements *continued*

Procedure

After completing each part of the experiment, record your observations in the appropriate data table. Recording data anywhere else increases the probability of recording an inaccurate value.

PART 1

1. Examine the centimeter scale of the plastic ruler. What are the smallest divisions?

_____

2. You can estimate a measurement to one-tenth of the smallest division on a measuring apparatus. To what fraction of a centimeter can you estimate with your plastic ruler?

_____

3. The uncertainty in a measurement is ± one-half of the smallest division. What is the uncertainty in a measurement made with your plastic ruler?

_____

4. Using the ruler, measure the inside diameter of the 100 mL graduated cylinder. Similarly, measure the inside height of the cylinder to the 50 mL mark. Record these measurements in **Data Table 1.**

Data Table 1	
Inside diameter of graduated cylinder	
Inside height of graduated cylinder	

Accuracy and Precision in Measurements *continued*

PART 2

5. Examine the gram scale of the balance. What are the smallest divisions?

_____

To what fraction of a gram can you estimate with a centigram balance?

_____

6. Examine the graduations on a 25 mL graduated cylinder, and determine the smallest fraction of a milliliter to which you could make a measurement. Does this match the *uncertainty* of a measurement made with a 100 mL graduated cylinder?

_____

_____

_____

_____

_____

7. Using the balance, measure the mass of the dry 25 mL cylinder. Record the mass in **Data Table 2**.

8. Fill the beaker half full of water, and measure its temperature to the nearest degree. Look up the density of water for this temperature in your textbook, and record in **Data Table 2** both the temperature and water density.

9. Fill your graduated cylinder with water to a level between 10 and 25 mL; accurately read and record the volume. Measure the mass of the water plus the cylinder, then record this value in **Data Table 2**. Save the water in the graduated cylinder for use in **Part 3**.

Data Table 2		
Mass of empty graduated cylinder		g
Temperature of water		°C
Density of water		g/cm^3
Volume of water		mL
Mass of graduated cylinder + water		g

Name _____ Class _____ Date _____

PART 3

10. Add a sufficient quantity of the assigned metal shot (aluminum, copper, or lead) to the cylinder containing the water (saved from **Part 2**) to increase the volume by at least 5 mL. Measure the volume and then the mass of the metal shot, water, and cylinder. Record your measurements in **Data Table 3**.

Data Table 3		
Volume of water (from **Part 2**)		mL
Mass of water + graduated cylinder (from **Part 2**)		g
Volume of metal + water		mL
Mass of metal + water + graduated cylinder		g

DISPOSAL

11. Clean up all apparatus and your lab station. Return equipment to its proper place. Wash your hands thoroughly with soap before you leave the lab and after all work is finished.

Analysis

Show all your calculations.

Part 1

1. Organizing Data Calculate the volume of the cylinder to the 50.0 mL graduation ($V = \pi \times r^2 \times h$).

2. Inferring Conclusions Assume the accepted value for the volume of the cylinder is 50.0 cm³. Calculate the experimental error and percentage error.

Part 2

3. Organizing Data Calculate the mass of water as measured by the balance.

Accuracy and Precision in Measurements *continued*

4. Organizing Data Calculate the mass of the water from its measured volume and its density ($m = D \times V$).

5. Inferring Conclusions Using the mass of water determined by the use of the balance as the *accepted value*, calculate the experimental error and percentage error in the mass determined using the volume and density.

Part 3

6. Organizing Data Determine the volume of the metal, using your measurement of the volume of water displaced by the metal.

7. Organizing Data Using your measurements in **Data Table 3,** determine the mass of the metal.

8. Organizing Data Calculate the density of the metal.

9. Inferring Conclusions In a handbook, your textbook, or on-line, look up the density of the metal you used. (In SI, the density of liquids and solids is equal to the specific gravity.) Calculate the experimental error and percentage error for the density of the metal shot you determined in item **8.**

Part 4

10. Organizing Conclusions Record in the table below five values obtained by you and your classmates for the density of the *same metal.*

Group number	Density (g/cm³)

Accuracy and Precision in Measurements *continued*

11. Evaluating Conclusions Calculate the average density (*D*) of the five results.

12. Inferring Conclusions Calculate the experimental error and the percentage error for the average density of the metal shot.

13. Evaluating Methods What value of a measurement must be known if the accuracy of an experimental measurement is to be determined?

_____

_____

14. Evaluating Methods What are the possible sources of experimental errors in this experiment?

_____

_____

Conclusions

1. Evaluating Conclusions Sarah and Jamal determined the density of a liquid three times. The values they obtained were 2.84 g/cm^3, 2.85 g/cm^3, and 2.80 g/cm^3. The accepted value is known to be 2.40 g/cm^3.

a. Are the values that Sarah and Jamal determined precise? Explain.

_____

_____

b. Are the values accurate? Explain.

_____

_____

 Skills Practice Experiment

Accuracy and Precision in Measurements *continued*

c. Calculate the percentage error for each density.

Skills Practice

Water of Hydration

Many ionic compounds, when crystallized from an aqueous solution, will take up definite amounts of water as an integral part of their crystal structures. You can drive off this water of crystallization by heating the hydrated substance to convert it to its anhydrous form. Because the law of definite composition holds for crystalline hydrates, the number of moles of water driven off per mole of the anhydrous compound is a simple whole number. If the formula of the anhydrous compound is known, you can use your data to determine the formula of the hydrate.

OBJECTIVES

Determine that all the water has been driven from a hydrate by heating a sample to constant mass.

Use experimental data to calculate the number of moles of water released by a hydrate.

Infer the empirical formula of the hydrate from the formula of the anhydrous compound and experimental data.

MATERIALS

- balance, centigram
- Bunsen burner and related equipment
- crucible and cover
- desiccator
- iron ring
- magnesium sulfate, Epsom salts, hydrated crystals, $MgSO_4 \cdot nH_2O$
- pipe-stem triangle
- ring stand
- sparker
- spatula
- crucible tongs

Always wear safety goggles and a lab apron to protect your eyes and clothing. If you get a chemical in your eyes, immediately flush the chemical out at the eyewash station while calling to your teacher. Know the locations of the emergency lab shower and the eyewash station and the procedures for using them.

Do not touch any chemicals. If you get a chemical on your skin or clothing, wash the chemical off at the sink while calling to your teacher. Make sure you carefully read the labels and follow the precautions on all containers of chemicals that you use. If there are no precautions stated on the label, ask your teacher what precautions you should follow. Do not taste any chemicals or items used in the laboratory. Never return leftovers to their original containers; take only small amounts to avoid wasting supplies.

Call your teacher in the event of a spill. Spills should be cleaned up promptly, according to your teacher's directions.

| Water of Hydration *continued*

When using a Bunsen burner, confine long hair and loose clothing. If your clothing catches on fire, WALK to the emergency lab shower and use it to put out the fire. Do not heat glassware that is broken, chipped, or cracked. Use tongs or a hot mitt to handle heated glassware and other equipment because hot glassware does not look hot.

Procedure

1. Throughout the experiment, handle the crucible and cover with clean crucible tongs only. Place the crucible and cover on the triangle as shown in **Figure A.** Position the cover slightly tipped, leaving only a small opening for any gases to escape. Preheat the crucible and its cover until the bottom of the crucible turns red.

CAUTION The crucible and cover are very hot after each heating. Remember to handle them only with tongs.

2. Using tongs, transfer the crucible and cover to a desiccator. Allow them to cool 5 min in the desiccator. Never place a hot crucible on a balance. When the crucible and cover are cool, determine their mass to the nearest 0.01 g. Record this mass in the **Data Table.**

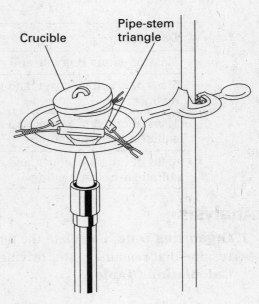

Crucible   Pipe-stem triangle

Figure A

3. Using a spatula, add approximately 5 g of magnesium sulfate hydrate crystals to the crucible. Determine the mass of the covered crucible and crystals to the nearest 0.01 g. Record this mass in the **Data Table.**

4. Place the crucible with the magnesium sulfate hydrate on the triangle, and again position the cover so that there is a small opening. Too large an opening may allow the hydrate to spatter out of the crucible. Heat the crucible very gently with a low flame to avoid spattering any of the hydrate. Increase the temperature gradually for 2 or 3 min. Then, heat strongly, but not red-hot, for at least 5 min.

5. Using tongs, transfer the crucible, cover, and contents to the desiccator, and allow them to cool for 5 min. Then, using the same balance you used in Step **2,** determine their mass. Be sure the crucible is sufficiently cool, because heat can affect your measurement. Record the mass in the **Data Table.**

Water of Hydration *continued*

6. Again heat the covered crucible and contents strongly for 5 min. Allow the crucible, cover, and contents to cool in the desiccator, and then use the same balance as before to determine their mass. If the last two mass measurements differ by no more than 0.01 g, you may assume that all the water has been driven off. Otherwise, repeat the heating process until the mass no longer changes. Record this constant mass in your **Data Table.**

DISPOSAL

7. Clean all apparatus and your lab station. Return equipment to its proper place. Dispose of the $MgSO_4$ in your crucible as your teacher directs. Do not pour any chemicals down the drain or in the trash unless your teacher directs you to do so. Wash your hands thoroughly after all work is finished and before you leave the lab.

Data Table	
Mass of empty crucible and cover	
Mass of crucible, cover, and magnesium sulfate hydrate	
Mass of crucible, cover, and anhydrous magnesium sulfate after 1st heating	
Mass of crucible, cover, and anhydrous magnesium sulfate after last heating	

Analysis

1. Organizing Data Calculate the mass of anhydrous magnesium sulfate (the residue that remained after driving off the water). Record the mass in the **Calculations Table.**

2. Organizing Data Calculate the number of moles of anhydrous magnesium sulfate. Record the number of moles in the **Calculations Table.**

3. Organizing Data Calculate the mass of water driven off from the hydrate. Record the mass in the **Calculations Table.**

Water of Hydration *continued*

4. Organizing Data Calculate the number of moles of water driven off from the hydrate. Record the number of moles in the **Calculations Table.**

5. Organizing Ideas Write the equation for the reaction that occurred when you heated hydrated $MgSO_4$ in this experiment. Use the letter n to represent the number of moles of water driven off per mole of anhydrous magnesium sulfate.

_____

6. Organizing Conclusions Using your answers to Calculations Items **2, 4,** and **5,** determine the mole ratio of $MgSO_4$ to H_2O to the nearest whole number. Record the ratio in the **Calculations Table.**

7. Organizing Conclusions Use your answer to Calculations Item **6** to write the formula for the magnesium sulfate hydrate. Record the formula in the **Calculations Table.**

_____

Calculations Table	
Mass of anhydrous magnesium sulfate	
Moles of anhydrous magnesium sulfate	
Mass of water driven off from hydrate	
Moles of water driven off from hydrate	
Mole ratio of anhydrous magnesium sulfate to water	
Empirical formula of the hydrate	

| Water of Hydration *continued* |

Conclusions

1. Applying Conclusions The following results were obtained when a solid was heated by three different lab groups. In each case, the students observed that when they began to heat the solid, drops of a liquid formed on the sides of the test tube.

Lab group	Mass before heating	Mass after heating
1	1.48 g	1.26 g
2	1.64 g	1.40 g
3	2.08 g	1.78 g

a. Could the solid be a hydrate? What evidence supports your answer?

_____

_____

_____

_____

b. If, after heating, the solid has a molar mass of 208 g/mol and a formula of XY, what is the formula of the hydrate? Use **Lab group 1**'s results.

2. Applying Conclusions Some cracker tins include a glass vial of drying material in the lid to keep the crackers crisp. In many cases, the material is a mixture of magnesium sulfate and cobalt chloride indicator. As the magnesium sulfate absorbs moisture ($MgSO_4 \cdot H_2O + 6H_2O \rightarrow MgSO_4 \cdot 7H_2O$), the indicator changes color from blue to pink ($CoCl_2 \cdot 4H_2O + 2H_2O \rightarrow CoCl_2 \cdot 6H_2O$). When this drying mixture becomes totally pink, it can be restored if it is heated in an oven. What two changes are caused by the heating?

_____

_____

_____

_____

_____

3. Applying Conclusions How does the experiment you performed exemplify the law of definite composition?

_____

_____

_____

_____

_____

_____

_____

_____

4. Analyzing Methods Why did you use the same balance each time you measured the mass of the crucible and its contents?

_____

_____

_____

5. Predicting Outcomes Use handbooks to investigate the properties of the compounds listed in the following table. Write answers to these questions in the table: Does the compound form a hydrate? How would you describe the appearance of the anhydrous compound? If your teacher approves, repeat the experimental procedure, using one of the listed compounds, and verify its hydrate formula. Explain any large deviation from the correct hydrate formula.

Name of Compound	Water of crystallization?	Appearance of anhydrous compound
Sodium carbonate		
Sodium sulfate		
Sodium aluminum sulfate		
Potassium chloride		
Magnesium chloride		
Copper sulfate		

Name _____ Class _____ Date _____

Evidence for Chemical Change

One way of knowing that a chemical change has occurred is to observe that the properties of the products are different from those of the reactants. A new product can then become a reactant in another chemical reaction. In this experiment, you will observe a sequence of changes that occur when a solution that begins as copper(II) nitrate is subjected to a series of different reactions. All of the reactions will take place in the same test tube. At each step, you will look for evidence that a new substance has formed as a result of a chemical change. You will also observe energy changes and relate them to chemical reactions.

OBJECTIVES

Observe evidence that a chemical change has taken place.

Infer from observations that a new substance has been formed.

Identify and record observations that show energy is involved in chemical change.

Observe the color and solubility of some substances.

Describe reactions by writing word equations.

MATERIALS

- aluminum wire, 12 cm
- beaker, 100 mL
- Bunsen burner and related equipment
- copper(II) nitrate, 1.0 M
- glass stirring rod
- gloves
- HCl, 1.0 M
- iron ring
- lab apron
- lab marker
- NaOH, 1.0 M
- ring stand
- ruler
- safety goggles
- test tube, 13 mm × 100 mm
- test-tube rack
- wire gauze, ceramic-centered

Always wear safety goggles and a lab apron to protect your eyes and clothing. If you get a chemical in your eyes, immediately flush the chemical out at the eyewash station while calling to your teacher. Know the location of the emergency lab shower and the eyewash station and the procedures for using them.

Do not touch any chemicals. If you get a chemical on your skin or clothing, wash the chemical off at the sink while calling to your teacher. Make sure you carefully read the labels and follow the precautions on all containers of chemicals that you use. If there are no precautions stated on the label, ask your teacher what precautions to follow. Do not taste any chemicals or items used in the laboratory. Never return leftover chemicals to their original containers; take only small amounts to avoid wasting supplies.

Evidence for Chemical Change *continued*

    Call your teacher in the event of a spill. Spills should be cleaned up promptly, according to your teacher's directions.

Acids and bases are corrosive. If an acid or base spills onto your skin or clothing, wash the area immediately with running water. Call your teacher in the event of an acid spill. Acid or base spills should be cleaned up promptly.

    Never put broken glass in a regular waste container. Broken glass should be disposed of separately according to your teacher's instructions.

    Do not heat glassware that is broken, chipped, or cracked. Use tongs or a hot mitt to handle heated glassware and other equipment because hot glassware does not always look hot.

    When using a Bunsen burner, confine long hair and loose clothing. If your clothing catches on fire, WALK to the emergency lab shower and use it to put out the fire.

    When heating a substance in a test tube, the mouth of the test tube should point away from where you and others are standing. Watch the test tube at all times to prevent the contents from boiling over.

Procedure

1. Put on safety goggles, gloves, and a lab apron.

2. Place 50 mL of water into the 100 mL beaker and heat it until boiling. This will be the water bath you will use in **Step 6.**

3. While the water bath is heating, use the lab marker and ruler to make three marks on the test tube that are 1 cm apart. Make the marks starting at the bottom of the test tube and moving toward the top.

4. Add 1.0 M copper(II) nitrate to the first mark on the test tube, as shown in **Figure 1.**

5. Add 1.0 M sodium hydroxide (NaOH) up to the second mark on the test tube, as shown in **Figure 2. CAUTION: Sodium hydroxide is corrosive. Be certain to wear safety goggles, gloves, and a lab apron. Avoid contact with skin and eyes. If any of this solution should spill on you, immediately flush the area with water and then notify your teacher.**

Mix the solutions with the stirring rod. Rinse the stirring rod thoroughly before setting it down on the lab table. Touch the bottom of the outside of the test tube to see if energy has been released. The copper-containing product in the test tube is copper(II) hydroxide. The other product is sodium nitrate. Record the changes that occur in the test tube in the space provided below.

Observations:

_____

_____

Evidence for Chemical Change *continued*

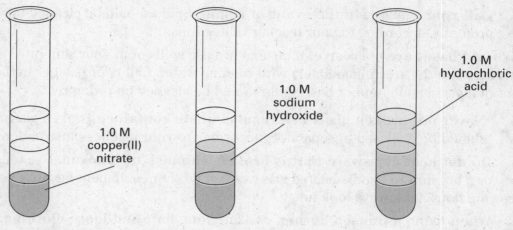

| **Figure 1** | **Figure 2** | **Figure 3** |

Figure 1: 1.0 M copper(II) nitrate

Figure 2: 1.0 M sodium hydroxide

Figure 3: 1.0 M hydrochloric acid

6. Put the test tube into the water bath you prepared in **Step 2**. Heat it until no more changes occur. The products of this reaction are copper(II) oxide and water. Record the changes that occur in the test tube.

Observations:

_____

_____

7. Remove the test tube from the hot-water bath. Turn off the burner. Cool the test tube and its contents for 2 min in room-temperature water. Add 1.0 M hydrochloric acid (HCl) to the third mark, as shown in **Figure 3**.

CAUTION: Hydrochloric acid is corrosive. Be certain to wear safety goggles, gloves, and a lab apron. Avoid contact with skin and eyes. Avoid breathing vapors. If any of this solution should spill on you, immediately flush the area with water and then notify your teacher.

Mix with the stirring rod. Rinse the stirring rod.

The new products are copper(II) chloride and water. Record the changes that occur in the test tube.

Observations:

_____

_____

_____

Evidence for Chemical Change *continued*

8. Place a 12-cm piece of aluminum wire in the test tube. Leave it until no reaction is observed. Touch the bottom of the test tube to check for temperature change. Two reactions take place. Copper(II) chloride and aluminum produce copper and aluminium chloride. The aluminum also reacts with the hydrochloric acid to form hydogen and aluminum chloride. Record the changes that occur in the test tube.

Observations:

_____

_____

_____

9. Remove the wire from the test tube. Compare the copper formed to a sample of copper wire. Record your observations.

Observations:

_____

_____

_____

10. Clean all apparatus and your lab station. Return equipment to its proper place. Dispose of chemicals and solutions in the containers designated by your teacher. Do not pour any chemicals down the drain or put them in the trash unless your teacher directs you to do so. Wash your hands thoroughly after all work is finished and before you leave the lab.

Analysis

1. **Organizing Ideas** What are some causes of chemical changes?

_____

_____

2. **Organizing Ideas** In what two ways is energy involved in chemical change? Cite some specific instances from this experiment.

_____

_____

_____

_____

Evidence for Chemical Change *continued*

3. Analyzing Information Identify all of the substances that are used or produced in this experiment. Distinguish between elements and compounds.

_____

_____

_____

4. Analyzing Methods In the last step of the experiment, where is the aluminum chloride? How could you recover it?

_____

_____

5. Inferring Conclusions What is the color of solutions of copper compounds?

_____

_____

_____

6. Analyzing Results Which substances involved in this experiment dissolve in water? Which do not dissolve?

_____

_____

_____

7. Organizing Conclusions Refer to the procedure steps in the experiment, and complete the following word equations.

a. copper(II) nitrate + sodium hydroxide $\rightarrow$

b. copper(II) hydroxide + energy $\rightarrow$

c. copper(II) oxide + hydrochloric acid $\rightarrow$

d. copper(II) chloride + aluminum $\rightarrow$

e. hydrochloric acid + aluminum $\rightarrow$

Evidence for Chemical Change *continued*

Conclusions

1. **Analyzing Information** List four types of observations that indicate when a chemical change has occurred.

   _____

   _____

2. **Relating Ideas** The chemical conversion of one product into another useful product is the basis for recycling. Explain how the type of reactions you observed in this experiment could be useful in the recycling of copper.

   _____

   _____

3. **Analyzing Conclusions** Describe the advantages and disadvantages of recycling metals as was done in this experiment.

   _____

   _____

   _____

Name _____ Class _____ Date _____

Boyle's Law

According to Boyle's law, the volume of a fixed amount of dry gas is inversely proportional to the pressure if the temperature is constant. Boyle's law may be stated mathematically as $P \propto \dfrac{1}{V}$ or $PV = k$ (where k is a constant). Notice that for Boyle's law to apply, two variables that affect gas behavior must be held constant: the amount of gas and the temperature.

 In this experiment, you will vary the pressure of air contained in a syringe and measure the corresponding change in volume. Because it is often impossible to determine relationships by just looking at data in a table, you will plot graphs of your results to see how the variables are related. You will make two graphs, one of volume versus pressure and another of the inverse of volume versus pressure. From your graphs you can derive the mathematical relationship between pressure and volume and verify Boyle's law.

OBJECTIVES

Determine the volume of a gas in a container under various pressures.

Graph pressure-volume data to discover how the variables are related.

Interpret graphs and verify Boyle's law.

MATERIALS

- Boyle's law apparatus or Vernier Gas Pressure Sensor and CBL setup
- carpet thread
- objects of equal mass (approximately 500 g each), 4

  Always wear safety goggles and a lab apron to protect your eyes and clothing. If you get a chemical in your eyes, immediately flush the chemical out at the eyewash station while calling to your teacher. Know the locations of the emergency lab shower and the eyewash station and the procedures for using them.

Procedure

1. Adjust the piston head of the Boyle's law apparatus so that it reads between 30 and 35 cm³. To adjust, pull the piston head all the way out of the syringe, insert a piece of carpet thread in the barrel, and position the piston head at the desired location, as shown in **Figure A** on the next page.
Note: Depending upon the Boyle's law apparatus that you use, you may find the volume scale on the syringe abbreviated in cc or cm³. Both abbreviations stand for cubic centimeter (1 cubic centimeter is equal to 1 mL).

2. While holding the piston in place, carefully remove the thread. Twist the piston several times to allow the head to overcome any frictional forces. Read the volume on the syringe barrel to the nearest 0.1 cm³. Record this value in your **Data Table** as the initial volume (for zero weights).

Boyle's Law *continued*

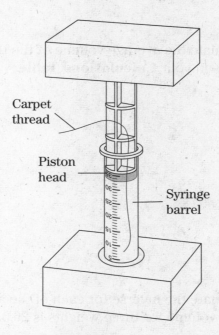

Carpet
thread

Piston
head

Syringe
barrel

Figure A

3. Place one of the weights on the piston. Give the piston several twists to overcome any frictional forces. When the piston comes to rest, read and record the volume to the nearest 0.1 cm^3.

4. Repeat step **3** for two, three, and four weights, and record your results in the **Data Table.**

5. Repeat steps **3** and **4** for at least two more trials. Record your results in the **Data Table.**

Data Table			
Pressure (number of weights)	Trial 1 volume (cm^3)	Trial 2 volume (cm^3)	Trial 3 volume (cm^3)
0			
1			
2			
3			
4			

DISPOSAL

6. Clean all apparatus and your lab station at the end of this experiment. Return equipment to its proper place.

Analysis

1. Organizing Data Calculate the average volume of the three trials for weights 0–4. Record your results in your **Calculations Table.**

2. Organizing Data Calculate the inverse for each of the average volumes. For example, if the average volume for three weights is 26.5 cm^3, then

$$\frac{1}{V} = \frac{1}{26.5 \text{ cm}^3} = \frac{0.0377}{\text{cm}^3}.$$ Record your results in your **Calculations Table.**

Calculations Table

Pressure (number of weights)	Average volume (cm^3)	1/Volume ($\times 10^{-2}$/cm^3)
0		
1		
2		
3		
4		

Boyle's Law *continued*

3. Analyzing Data Plot a graph of volume versus pressure on the grid below. Because the number of weights added to the piston is directly proportional to the pressure applied to the gas, you can use the number of weights to represent the pressure. Notice that pressure is plotted on the horizontal axis and volume is plotted on the vertical axis. Draw the smoothest curve that goes through most of the points.

Volume vs. Pressure

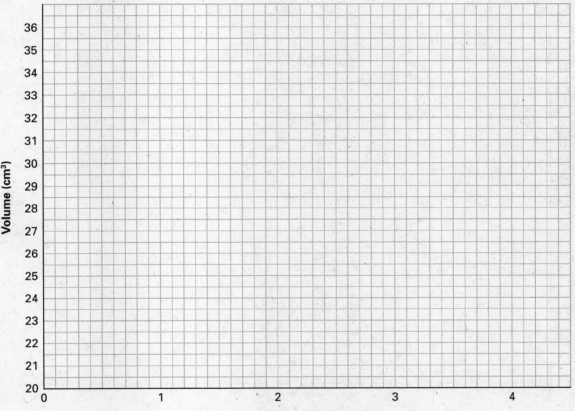

4. Interpreting Graphics Does your graph indicate that a change in volume is directly proportional to a change in pressure? Explain your answer.

_____

_____

_____

_____

5. Analyzing Data On the grid below, plot a graph of $\dfrac{1}{volume}$ versus *pressure*.

Notice that *pressure* is on the horizontal axis and $\dfrac{1}{volume}$ is on the vertical

axis. Draw the best line that goes through the majority of the points.

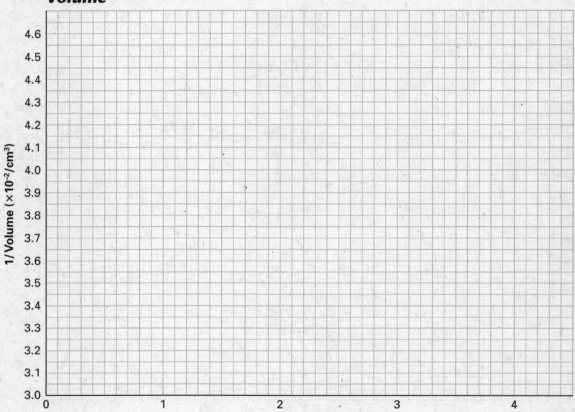

$\dfrac{1}{\textbf{Volume}}$ **vs. *Pressure***

6. Interpreting Graphics What do you conclude about the mathematical relationship between the pressure applied to a gas and its corresponding volume?

_____

_____

_____

_____

Conclusions

1. **Applying Models** Correlate the observed relationship between the pressure and volume of a gas with the kinetic theory description of a gas.

   _____

   _____

   _____

2. **Applying Conclusions** Use your graph of $\dfrac{1}{Volume}$ vs. *Pressure* to predict the volume of the gas if 2.5 weights were used.

   _____

   _____

3. **Analyzing Conclusions** If a normal sea-level recipe were used to prepare a cake at a location 1000 m below sea level, the cake would be much flatter than expected. Explain why, and offer a solution. (Hint: Consider how barometric pressure differs at this altitude and what effect that might have on the ability of the cake to rise.)

   _____

   _____

   _____

   _____

   _____

Skills Practice

Molar Volume of a Gas

Magnesium is an active metal that readily reacts with hydrochloric acid to produce hydrogen gas. In this experiment, you will react a known amount of magnesium, Mg, with hydrochloric acid, HCl, and collect and measure the volume of hydrogen gas, H_2, produced. From the volume of the gas measured at atmospheric pressure and the temperature of the lab, you can calculate the volume of the gas at standard temperature and pressure (STP). Knowing the mass of magnesium, you can calculate the number of moles of magnesium consumed and determine the volume of hydrogen at STP produced by the reaction of 1 mol of magnesium. This conclusion can then be related to the balanced equation for the reaction.

OBJECTIVES

Determine the volume of hydrogen produced by the reaction of a known mass of magnesium with hydrochloric acid.

Compute the volume of the gas at standard temperature and pressure.

Relate the volume of gas to the number of moles of magnesium reacted.

Infer the volume of one mole of gas at standard temperature and pressure.

MATERIALS

- 6 M HCl
- 50 mL beaker
- 50 mL eudiometer
- 400 mL beaker
- 1000 mL graduated cylinder or hydrometer jar

- buret clamp
- magnesium, Mg, ribbon (untarnished)
- ring stand
- rubber stopper, one-hole, #00
- thermometer, nonmercury, 0-100°C
- thread

Always wear safety goggles and a lab apron to protect your eyes and clothing. If you get a chemical in your eyes, immediately flush the chemical out at the eyewash station while calling to your teacher. Know the locations of the emergency lab shower and the eyewash station and the procedures for using them.

Do not touch any chemicals. If you get a chemical on your skin or clothing, wash the chemical off at the sink while calling to your teacher. Make sure you carefully read the labels and follow the precautions on all containers of chemicals that you use. If there are no precautions stated on the label, ask your teacher what precautions to follow. Do not taste any chemicals or items used in the laboratory. Never return leftover chemicals to their original containers; take only small amounts to avoid wasting supplies.

 Call your teacher in the event of a spill. Spills should be cleaned up promptly, according to your teacher's directions.

Never put broken glass in a regular waste container. Broken glass should be disposed of separately according to your teacher's instructions.

Procedure

1. Fill a 400 mL beaker two-thirds full of water that has been adjusted to room temperature.

2. Measure about 4.4 cm of magnesium ribbon to the nearest 0.1 cm. Your piece of magnesium should not exceed 4.5 cm. Record the length of the ribbon in the **Data Table.**

3. Obtain the mass of 1 m of magnesium ribbon from your teacher and record this mass in the **Data Table.**

4. Roll the length of the magnesium ribbon into a loose coil. Tie it tightly with one end of a piece of thread approximately 25 cm in length. All the loops of the coil should be tied together as shown in **Figure A.**

5. This step requires the use of 6 M hydrochloric acid and will be performed by your teacher.
   CAUTION Hydrochloric acid is caustic and corrosive. Avoid contact with skin and eyes. Avoid breathing the vapor. Make certain that you are wearing safety goggles, a lab apron, and gloves when working with the acid. If any acid should splash on you, immediately flush the area with water and then report the incident to your teacher. If you should spill any acid on the counter top or floor, ask your teacher for cleanup instructions.

   Your teacher will carefully pour approximately 10 mL of 6 M HCl into a 50 mL beaker and then pour the 10 mL of HCl into your gas-measuring tube, or eudiometer.

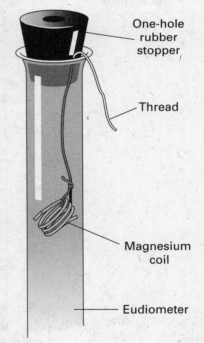

One-hole rubber stopper

Thread

Magnesium coil

Eudiometer

Figure A

6. While holding the eudiometer in a slightly tipped position, very slowly pour water from the 400 mL beaker into the eudiometer, being careful to layer the water over the acid so that they do not mix. Add enough water to fill the eudiometer to the brim.

7. Lower the magnesium coil into the water in the eudiometer tube to a depth of about 5 cm. Insert the rubber stopper into the eudiometer to hold the thread in place, as shown in **Figure A.** The one-hole stopper should displace some water from the tube. This ensures that no air is left inside the tube.

Name _____ Class _____ Date _____

Molar Volume of a Gas *continued*

8. Cover the hole of the stopper with your finger, and invert the eudiometer into the 400 mL beaker of water. Clamp the eudiometer tube into position on the ring stand, as shown in **Figure B.** The acid flows down the tube (why?) and reacts with the magnesium. Is the acid now more concentrated or dilute? Answer these questions and describe your observations in the place provided following the **Data Table.**

9. When the magnesium has disappeared entirely and the reaction has stopped, cover the stopper hole with a finger and carefully transfer the eudiometer tube to a 1000 mL graduated cylinder or other tall vessel that has been filled with water. Adjust the level of the eudiometer tube in the water as shown in **Figure C.** The levels of the liquid inside the eudiometer and the graduated cylinder should be the same. The hydrogen gas produced by the reaction will move to the top of the eudiometer. Read the volume of hydrogen gas you collected as accurately as possible.

10. Read the thermometer and record the temperature of the room and the hydrogen gas, H_2 in the **Data Table.** Determine the atmospheric pressure and record it in the **Data Table.**

11. Use the table of water vapor pressures in your textbook to find the vapor pressure of water at the temperature of the room. Record this water vapor pressure in the **Data Table.**

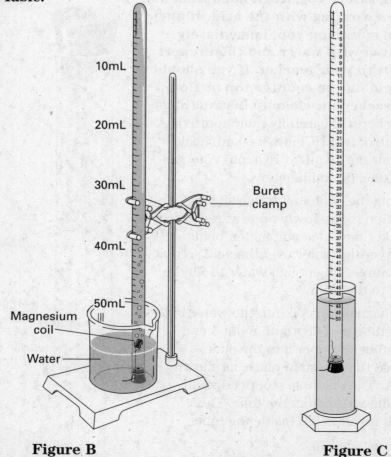

Figure B   **Figure C**

Molar Volume of a Gas *continued*

Disposal

12. Clean all apparatus and your lab station. Return equipment to its proper place. Dispose of chemicals and solutions in the containers designated by your teacher. Do not pour any chemicals down the drain or in the trash unless your teacher directs you to do so. Wash your hands thoroughly before you leave the lab and after all work is finished.

Data Table		
Length of Mg used	**4.2**	cm
Mass per meter of Mg	**0.981**	g/m
Volume of H_2 collected under lab conditions	**42.6**	mL
Temperature of H_2 collected	**20.0**	°C
Atmospheric pressure	**755**	mm Hg
Vapor pressure of water at observed temperature	**17.5**	mm Hg

Observations:

_____

_____

_____

_____

_____

Analysis

For the following questions, show all your calculations in the space provided.

1. **Organizing Data** From the length of the magnesium ribbon you used and the mass of a meter of magnesium ribbon, determine the mass of the magnesium that reacted.

2. **Organizing Data** Calculate the number of moles of magnesium that reacted. (The molar mass of Mg is 24.3 g/mol.)

| Molar Volume of a Gas *continued*

3. **Organizing Data** Because the hydrogen gas was collected over water, two gases were present in the eudiometer: hydrogen and water vapor. Calculate the partial pressure of the hydrogen gas you collected at the temperature of the room. (Hint: The sum of the two partial pressures equals atmospheric pressure.)

4. **Organizing Data** Calculate the volume of the dry hydrogen you collected at room temperature and standard pressure (760 mm Hg). (Hint: $P_1V_1 = P_2V_2$)

5. **Organizing Data** Calculate the volume of dry hydrogen gas at standard temperature (273 K). $\left(\text{Hint: } \dfrac{V_1}{T_1} = \dfrac{V_2}{T_2}\right)$

6. **Relating Ideas** Write a balanced equation for the reaction of magnesium with HCl. The products are hydrogen gas and $MgCl_2$.

7. **Inferring Conclusions** From the balanced equation, calculate the volume of dry hydrogen gas that would have been produced had you used 1 mol of magnesium at standard temperature and pressure.

8. **Analyzing Methods** Why was it necessary to make a water vapor pressure correction of the atmospheric pressure in this experiment?

_____

_____

_____

_____

_____

Molar Volume of a Gas *continued*

9. Analyzing Methods Why was it necessary to adjust the level of the eudiometer in the cylinder, as in **Figure C,** so that the level of water in the
eudiometer was the same as the level of water in the cylinder?

_____

_____

_____

_____

_____

_____

_____

_____

_____

Conclusions

1. Applying Conclusions From the balanced equation in Analysis question 6,
determine the volume of hydrogen gas at standard temperature and pressure
that can be produced from 3 mol of magnesium reacting with hydrochloric
acid.

_____

_____

_____

_____

Molar Volume of a Gas *continued*

2. **Applying Conclusions** Since the 1930s, alloys of aluminum and magnesium have been used in the manufacture of pots and pans. An unanswered question is whether a small amount of these metals that might dissolve in food as it cooks is beneficial, harmful, or of no consequence. What are some examples of foods that will react with aluminum/magnesium pans?

_____

_____

_____

_____

_____

Skills Practice

A Close Look at Soaps and Detergents

Soap is a surface-active agent (surfactant) that lowers the surface tension of water, allowing fat or oil-bearing soil particles to be suspended in water (emulsification). Although soaps are excellent cleansers in soft water, they are ineffective in hard-water conditions. Hard water contains aqueous salts of magnesium, calcium, and iron. When soap is used in hard water, the insoluble calcium salts of the fatty acids and other precipitates are deposited as curds. This precipitate is commonly referred to as bathtub ring or scale. To overcome this problem, inexpensive synthetic detergents were developed around 1950. Although immensely popular, these compounds were not biodegradable and were replaced in the mid-1960s.

Synthetic detergents, like soaps, have a long hydrocarbon chain labeled as a tail that is attached to a hydrophilic head. Synthetic surfactants are classified as anionic, cationic, or nonionic, depending on the type of hydrophilic head. The hydrophilic portion of a synthetic detergent is often a sulfonate, phosphate, or something other than the carboxyl group that is found in the soap. Therefore, synthetic surfactants are effective in both hard and soft water. No precipitates form when calcium, magnesium, or iron ions are present in solution. The most widely used group of synthetic detergents is the linear alkyl sulfonates (LAS). LAS are straight-chain compounds that have 10 or more carbon atoms and are easily degraded by bacteria.

Because water is a polar substance, it cannot remove dirt from fabrics if the dirt is suspended in oil and grease, which are nonpolar. The hydrocarbon tail of a detergent molecule dissolves itself in an oily substance but leaves the polar head outside the oily surface. Many detergent molecules continue to orient themselves in this way until the dirt containing oil is encapsulated. The oil droplet is then lifted away from the fabric and suspended in the water as a droplet or micelle.

OBJECTIVES

Observe the action of surfactants in reducing surface tension.

Evaluate the foaming capacity of commercial soaps and synthetic detergents.

Determine the efficiency of soaps and synthetic detergents in emulsifying a fat sample in water.

| A Close Look at Soaps and Detergents *continued*

MATERIALS

- alcohol solution of Sudan IV, 0.5%
- balance
- beaker, 500 mL
- Ca(NO$_3$)$_2$ solution, 4%
- detergent (powdered and liquid)
- detergent solution, 1.5%
- Erlenmeyer flask, 125 mL (5)
- fabric, (two 2-in. squares)
- graduated cylinder, 100 mL
- hot plate

- lard
- medicine dropper
- metric ruler
- No. 5 solid rubber stoppers, 5
- soap (bar and liquid)
- stirring rod
- thermometer
- wax paper
- wax pencil

Always wear safety goggles and a lab apron to protect your eyes and clothing. If you get a chemical in your eyes, immediately flush the chemical out at the eyewash station while calling to your teacher. Know the location of the emergency lab shower and eyewash station and the procedures for using them.

Do not touch any chemicals. If you get a chemical on your skin or clothing, wash the chemical off at the sink while calling to your teacher. Make sure you carefully read the labels and follow the precautions on all containers of chemicals that you use. If there are no precautions stated on the label, ask your teacher what precautions to follow. Do not taste any chemicals or items used in the laboratory. Never return leftovers to their original container; take only small amounts to avoid wasting supplies.

 Call your teacher in the event of a spill. Spills should be cleaned up promptly, according to your teacher's directions.

Never put broken glass in a regular waste container. Broken glass should be disposed of separately according to your teacher's instructions.
   Never stir with a thermometer because the glass around the bulb is fragile and might break.

A Close Look at Soaps and Detergents *continued*

Procedure

PART 1–INVESTIGATING THE ACTION OF A SURFACE-ACTING AGENT (SURFACTANT)

1. Put on safety goggles, gloves, and lab apron.

2. Using a medicine dropper, carefully add single drops of distilled water to a square of clean fabric. Describe what occurs.

_____

_____

_____

_____

3. Using a second medicine dropper, carefully add drops of detergent solution on top of the beaded water drops. Describe what happens.

_____

_____

_____

_____

PART 2–EVALUATING CLEANSING CHARACTERISTICS OF SOAPS AND SYNTHETIC DETERGENTS

4. Use a wax pencil to label a set of five 5 Erlenmeyer flasks 1, 2, 3, 4, and 5.

5. Obtain five soap solutions from your teacher. Record in the **Table 1** the manufacturer's brand name next to each sample number. Then add 10 mL of each soap or detergent sample to its respective numbered flask.

6. Stopper flask 1 securely with a rubber stopper. Hold the flask with your thumb on the stopper, and shake it vigorously for 15 seconds. Allow the solution to stand for 30 seconds. Observe and measure the level of the foam. Record this value in the **Table 1.** Repeat this process for flasks 2 through 5. Which product has the highest foam level?

_____

_____

_____

_____

| A Close Look at Soaps and Detergents *continued*

7. To each flask add 4 drops of 4% $Ca(NO_3)_2$ solution from a medicine dropper. Stopper the flask, and with your thumb on the stopper, shake vigorously for 15 seconds. Allow the solution to stand for 30 seconds. Observe and measure the level of foam. Record this value in the **Table 1.** Describe how $Ca(NO_3)_2$ affects a surfactant's ability to foam.

_____

_____

_____

_____

_____

PART 3–EVALUATING THE EFFECTIVENESS OF SOAPS AND SYNTHETIC DETERGENTS IN EMULSIFYING FAT

8. Fill a 500 mL beaker with 250 mL of distilled water. Set the beaker on a hot plate and heat the water to 55°C. Try to maintain this temperature as closely as possible throughout the experiment.

9. Measure a 0.1 g sample of colored fat onto a piece of wax paper.

10. Add the fat sample to the water in the beaker. Use a glass stirring rod to swirl the water and fat. Describe what happens.

_____

_____

_____

11. Fill a graduated cylinder with exactly 100 mL of test sample 1.

12. While stirring constantly, slowly add sample 1 to the water/lard mixture until the fat is completely emulsified, as evidenced by the dispersion of the colored fat in water. Record in the **Table 1** the volume of detergent necessary to completely emulsify the lard (fat) globule. Clean the beaker and the graduated cylinder. Repeat steps 8–12 for each of the remaining samples.

13. Identify each sample tested as either a detergent or a soap in the **Table 1.** Based upon emulsification data, which soap and synthetic detergent brand is the most effective emulsifier? Give reasons for your answer.

_____

_____

_____

_____

A Close Look at Soaps and Detergents *continued*

TABLE 1: COMPARING SOAP AND DETERGENT BRANDS

Sample number	Foam level in water (cm)	Foam level in 4% Ca(NO₃)₂ (cm)	Volume needed to emulsify 0.1 g of fat (mL)	Efficiency rating (g/mL)	Soap or detergent
1					
2					
3					
4					
5					
6					

DISPOSAL

14. Clean all apparatus and your lab station. Return equipment to its proper place. Dispose of chemicals and solutions in the containers designated by your teacher. Do not pour any chemicals down the drain or put them in the trash unless your teacher directs you to do so. Wash your hands thoroughly after all work is finished and before you leave the lab.

Analysis

1. Organizing Data For each sample tested, calculate its relative efficiency as a dirt and grease remover by dividing the 0.1 g of fat emulsified by the volume in milliliters of detergent added to the fat/water mixture.

A Close Look at Soaps and Detergents *continued*

2. Organizing Data Construct a bar graph comparing the efficiency of soaps and detergents in emulsifying fat. Place surfactant samples along the *x*-axis and the efficiency along the *y*-axis.

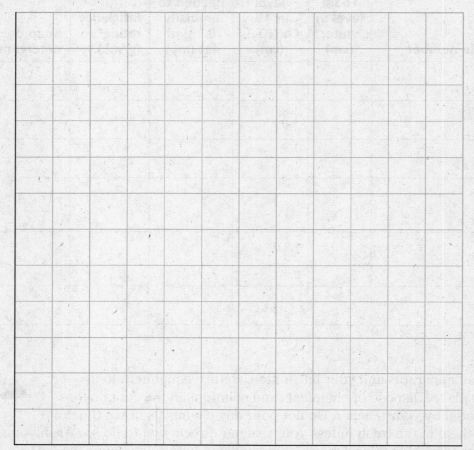

3. Applying Conclusions You are demonstrating the effect of hard water on sudsing. You have all the demonstration materials except a source of hard water. All that is available to you is milk or a roll of antacid tablets. Can the demonstration proceed? Justify your answer.

_____

_____

_____

_____

A Close Look at Soaps and Detergents *continued*

4. Analyzing Data and Inferring Conclusions According to entries in the
Table 1, which would be more efficient at dirt removal: a liquid detergent or
a solid detergent? Explain.

_____

_____

_____

_____

Conclusions

1. Inferring Conclusions Many product labels on detergents state that the
product will not harm septic systems. Suggest a reason for this based on your
foam data.

_____

_____

_____

2. Inferring Conclusions Visit your local grocer and look for a popular brand of
colorless hand gel. Write down the ingredients and compare them with the
ingredients in various shampoos. Summarize your conclusions.

_____

_____

_____

Name _____ Class _____ Date _____

Household Indicators

A visual indicator is a chemical substance that reflects the nature of the chemical system in which it is placed by changing color. Most visual indicators are complex organic molecules that exist in multiple colored forms, one of which could be colorless, depending on the chemical environment. Many visual indicators are used to test a solution's acidity.

Acid-base indicators respond to hydronium ion concentrations, $[H_3O^+]$. Acidic solutions have an excess of H_3O^+ ions, while basic or alkaline solutions have few H_3O^+ ions. A measure of the $[H_3O^+]$ is pH. Chemists use p to mean "power," so pH means the power of the hydronium ions. Formally, pH is defined as the negative logarithm of the $[H_3O^+]$ of a solution.

The most common acid-base indicator is litmus, a blue coloring matter extracted from various species of lichens. The chief component of litmus is azolitmin. Widely distributed among the higher plants, anthocyanins constitute most of the yellow, red, and blue colors in flowers and fruits. Anthocyanins are excellent acid-base indicators because they exhibit color changes over a wide range of pH values. Red cabbage is a ready source of this pigment. Unlike litmus and anthocyanins, universal indicator is a mixture of various synthetic indicator molecules. Universal indicator provides pH scale coverage from 1 to 14 pH units.

In this investigation, you will extract the pigment from red-cabbage leaves and use it to prepare strips of pH indicator paper. You will use these paper strips to test the pH of common household materials. Then you will compare your results with pH tests made with litmus and universal pH paper.

OBJECTIVES

Extract anthocyanins from red cabbage leaves.

Prepare a pH indicator paper with the red cabbage anthocyanins.

Construct a color indicator chart for anthocyanins from red cabbage.

Compare and evaluate the accuracy of indicator papers for recording pH values of common items.

Household Indicators *continued*

MATERIALS

- 6-well spot plate, white
- beaker, 250 mL (2)
- CH_3COOH, 0.1 M
- clothespins
- colored pencils (red, rose, purple, blue, green, yellow)
- dropper bottles for acid solutions
- filter paper, 9 cm diameter
- fine-point scissors
- forceps
- H_3BO_3, 0.1 M
- HCl, 0.1 M
- hot mitt or pot holder
- hot plate
- household ammonia, 0.1%

- household materials: ammonia, apple, distilled water, fresh egg, grapefruit, fruit jelly, ginger ale, lemon, milk, milk of magnesia, molasses, orange, mineral water, sauerkraut, sweet potato, tomato, salt water, and tap water
- litmus paper, neutral
- medicine dropper
- metric ruler
- $NaHCO_3$, 0.1 M
- NaOH, 0.1 M
- red cabbage
- string
- universal pH paper, (1–14 pH units)
- wax pencil

Always wear safety goggles, gloves, and a lab apron to protect your eyes and clothing. If you get a chemical in your eyes, immediately flush the chemical out at the eyewash station while calling to your teacher. Know the location of the emergency lab shower and eyewash station and the procedures for using them.

Do not touch any chemicals. If you get a chemical on your skin or clothing, wash the chemical off at the sink while calling to your teacher. Make sure you carefully read the labels and follow the precautions on all containers of chemicals that you use. If there are no precautions stated on the label, ask your teacher what precautions to follow. Do not taste any chemicals or items used in the laboratory. Never return leftovers to their original container; take only small amounts to avoid wasting supplies.

Call your teacher in the event of a spill. Spills should be cleaned up promptly, according to your teacher's directions.

Acids and bases are corrosive. If an acid or base spills onto your skin or clothing, wash the area immediately with running water. Call your teacher in the event of an acid spill. Acid or base spills should be cleaned up promptly.

Never put broken glass in a regular waste container. Broken glass should be disposed of separately according to your teacher's instructions.

⬥ **Do not heat glassware that is broken, chipped, or cracked.** Use tongs or a hot mitt to handle heated glassware and other equipment because hot glassware does not always look hot.

Procedure

PART 1–EXTRACTING ANTHOCYANINS FROM RED CABBAGE

1. Put on safety goggles, gloves, and lab apron.

2. Choose a red-cabbage leaf that has a dark purple color. Tear the leaf into small pieces.

3. Fill a 250 mL beaker 2/3 full with the leaf pieces. Add enough distilled water to just cover the pieces of cabbage leaf. Place the beaker on a hot plate, and bring the water to a slow boil. Continue heating for 5 minutes. Then turn off the heat, and allow the mixture to cool for 10 to 15 minutes.

PART 2–DYEING INDICATOR STRIPS

4. Using a hot mitt, remove the beaker from the hot plate. Carefully decant the cooled purple liquid into a second 250 mL beaker. Dispose of the cabbage-leaf material as directed by your teacher.

5. Using a pencil, write your initials on six filter paper circles. Using forceps, submerge the filter paper into the beaker containing the cabbage extract. Make sure that the papers are thoroughly wet. Then, using forceps, remove the filter papers from the beaker and allow them to dry thoroughly.

6. After drying, use scissors to cut from the filter paper 30 individual strips, each measuring 1 cm × 6 cm.

PART 3–CONSTRUCTING A COLOR INDICATOR CHART FOR THE ANTHOCYANIN-DYED pH PAPER

7. Use a wax pencil to label six wells in a spot plate 1 to 6.

8. Following the table below, add 10 drops of each named acid or base solution to its labeled well.

1	2	3	4	5	6
0.1 M CH_3COOH	0.1% household ammonia	0.1 M H_3BO_3	0.1 M HCl	0.1 M $NaHCO_3$	0.1 M NaOH

9. Into each of the six wells, dip a separate strip of the indicator paper prepared in **Part 2**. Record the color in **Table 1**. Rinse off the well plates.

| Household Indicators *continued*

TABLE 1: pH AND INDICATOR COLOR

pH	Substance	Indicator color
1.0	0.1 M hydrochloric acid, HCl	
2.9	0.1 M acetic acid, CH_3COOH	
5.2	0.1 M boric acid, H_3BO_3	
8.4	0.1 M sodium bicarbonate, $NaHCO_3$	
11.1	0.1 M ammonia, NH_3	
13.0	0.1 M sodium hydroxide, NaOH	

10. Use colored pencils and the entries in **Table 1** to fill in the following color indicator chart.

pH color chart for anthocyanins extracted from red caggage

1	2	3	4	5	6	7	8	9	10	11	12	13	14

PART 4–COMPARING AND EVALUATING THE ACCURACY OF INDICATOR PAPERS

11. Test each item listed in **Table 2** with universal pH paper, litmus paper, and the anthocyanin paper prepared in **Part 2**. Test each of the liquids listed by placing 10 drops of the liquid in a spot-plate well. Test the juice of each solid. Record the pH values in **Table 2.**

12. Clean all apparatus and your lab station. Return equipment to its proper place. Dispose of chemicals and solutions in the containers designated by your teacher. Do not pour any chemicals down the drain or put them in the trash unless your teacher directs you to do so. Wash your hands thoroughly after all work is finished and before you leave the lab.

TABLE 2: pH OF COMMON SUBSTANCES

Common substance	Measured pH	pH (neutral litmus paper)	pH (anthocyanin dyed paper)	pH (universal pH paper)
Ginger ale	2.0–4.0			
Lemon slice (lemon juice)	2.2–2.4			
Apple cider	2.4–2.9			
Apple slice	2.9–3.3			
Grapefruit (cut)	3.0–3.3			
Jellies, fruit	3.0–3.3			
Orange slice	3.0–4.0			
Sauerkraut	3.4–3.6			
Tomato slice	4.1–4.4			
Molasses	5.0–5.4			
Sweet potato slice	5.3–5.6			
Mineral water	6.2–9.4			
Milk	6.4–6.8			
Tap water	6.5–8.0			
Distilled water (CO_2-free)	7.0			
Fresh egg	7.6–8.0			
Salt water	8.0–8.4			
Milk of Magnesia	10.5			
Household ammonia	10.5–11.9			

Analysis

1. **Organizing Conclusions** Do all three pH test papers cover the entire pH scale range? Use your data to support your answer.

_____

_____

_____

Name _____ Class _____ Date _____

Household Indicators *continued*

2. Analyzing Data Which pH test paper was the most accurate within its respective pH scale range? Explain your reasoning.

_____

_____

_____

_____

3. Inferring Conclusions Suggest reasons for the variation in measured pH values.

_____

_____

_____

_____

_____

4. Analyzing Data Use the pH of a fresh egg to determine the hydronium ion concentration for the egg.

_____

_____

Conclusions

1. Inferring Conclusions The pH of cow's milk is approximately 6.5. When milk spoils, we sometimes say it has gone sour because of the lemon-like taste it develops. From your knowledge of food pH chemistry, what is happening to the milk's pH?

_____

_____

_____

_____

_____

_____

2. Designing Experiments The acid-base indicator, phenolphthalein, is also a mild laxative and can be the active ingredient in commercial chocolate-flavored laxatives. Phenolphthalein is not soluble in water, but it is soluble in rubbing alcohol (70% isopropyl alcohol). How would you evaluate its effectiveness as an indicator?

_____

_____

_____

_____

_____

_____

Skills Practice

Titration with an Acid and a Base

Titration is a process in which you determine the concentration of a solution by measuring what volume of that solution is needed to react completely with a standard solution of known volume and concentration. The process consists of the gradual addition of the standard solution to a measured quantity of the solution of unknown concentration until the number of moles of hydronium ion, H_3O^+, equals the number of moles of hydroxide ion, OH^-. The point at which equal numbers of moles of acid and base are present is known as the equivalence point. An indicator is used to signal when the equivalence point is reached. The chosen indicator must change color at or very near the equivalence point. The point at which an indicator changes color is called the end point of the titration. Phenolphthalein is an appropriate choice for this titration. In acidic solution, phenolphthalein is colorless, and in basic solution, it is pink.

At the equivalence point, the number of moles of acid equals the number of moles of base.

$$(1) \quad \text{moles of } H_3O^+ = \text{moles of } OH^-$$

By definition

$$(2) \quad molarity \text{ (mol/L)} = \frac{moles}{volume} \text{ (L)}$$

If you rearrange equation **2** in terms of moles, equation **3** is obtained.

$$(3) \quad moles = molarity \text{ (mol/L)} \times volume \text{ (L)}$$

When equations **1** and **3** are combined, you obtain the relationship that is the basis for this experiment, assuming a one-to-one mole ratio and the units of volume are the same for both the acid and base.

$$(4) \quad \text{molarity of acid} \times \text{volume of acid} = \text{molarity of base} \times \text{volume of base}$$

In this experiment, you will be given a standard hydrochloric acid, HCl, solution and told what its concentration is. You will carefully measure a volume of it and determine how much of the sodium hydroxide, NaOH, solution of unknown molarity is needed to neutralize the acid sample. Using the data you obtain and equation **4**, you can calculate the molarity of the NaOH solution.

OBJECTIVES

Use burets to accurately measure volumes of solution.

Recognize the end point of a titration.

Describe the procedure for performing an acid-base titration.

Determine the molarity of a base.

| Titration with an Acid and a Base *continued*

MATERIALS

- 0.500 M HCl
- 50 mL burets, 2
- 100 mL beakers, 3
- 125 mL Erlenmeyer flask
- double buret clamp

- NaOH solution of unknown molarity
- phenolphthalein indicator
- ring stand
- wash bottle filled with deionized water

 Always wear safety goggles, gloves, and a lab apron to protect your eyes and clothing. If you get a chemical in your eyes, immediately flush the chemical out at the eyewash station while calling to your teacher. Know the locations of the emergency lab shower and eyewash station and the procedures for using them.

Do not touch any chemicals. If you get a chemical on your skin or clothing, wash the chemical off at the sink while calling to your teacher. Make sure you carefully read the labels and follow the precautions on all containers of chemicals that you use. If there are no precautions stated on the label, ask your teacher what precautions to follow. Do not taste any chemicals or items used in the laboratory. Never return leftovers to their original container; take only small amounts to avoid wasting supplies.

Call your teacher in the event of a spill. Spills should be cleaned up promptly, according to your teacher's directions.

Never put broken glass into a regular waste container. Broken glass should be disposed of properly.

Procedure

1. Set up the apparatus as shown in **Figure A.** Label the burets *NaOH* and *HCl.* Label two beakers *NaOH* and *HCl.* Place approximately 80 mL of the appropriate solution into each beaker.

2. Pour 5 mL of NaOH solution from the beaker into the NaOH buret. Rinse the walls of the buret thoroughly with this solution. Allow the solution to drain through the stopcock into another beaker and discard it. Rinse the buret two more times in this manner, using a new 5 mL portion of NaOH solution each time. Discard all rinse solutions.

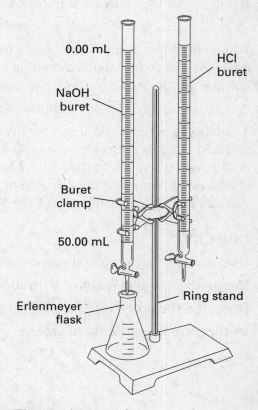

Figure A

Titration with an Acid and a Base *continued*

3. Fill the buret with NaOH solution to above the zero mark. Withdraw enough solution to remove any air from the buret tip, and bring the liquid level down within the graduated region of the buret.

4. Repeat steps **2** and **3** with the HCl buret, using HCl solution to rinse and fill it.

5. For trial 1, record the initial reading of each buret, estimating to the nearest 0.01 mL, in the **Data Table.** For consistent results, have your eyes level with the top of the liquid each time you read the buret. Always read the scale at the bottom of the meniscus.

6. Draw off about 10 mL of HCl solution into an Erlenmeyer flask. Add some deionized water to the flask to increase the volume. Add one or two drops of phenolphthalein solution as an indicator.

7. Begin the titration by slowly adding NaOH from the buret to the Erlenmeyer flask while mixing the solution by swirling it, as shown in **Figure B.** Stop frequently, and wash down the inside surface of the flask, using your wash bottle.

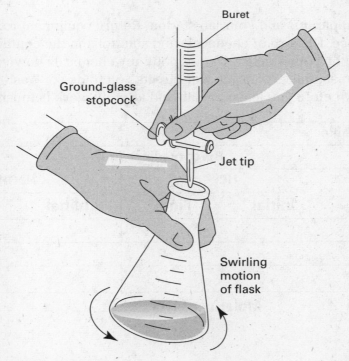

Figure B

69
Skills Practice Experiment

| Titration with an Acid and a Base *continued*

8. When the pink color of the solution begins to appear and linger at the point of contact with the base, add the base drop by drop, swirling the flask gently after each addition. When the last drop added causes the pink color to remain throughout the whole solution and the color does not disappear, stop the titration. A white sheet of paper under the Erlenmeyer flask makes it easier to detect the color change.

9. Add HCl solution dropwise just until the pink color disappears. Add NaOH again, dropwise, until the pink color remains. Go back and forth over the end point several times until one drop of the basic solution just brings out a faint pink color. Wash down the inside surface of the flask, and make dropwise addition of NaOH, if necessary, to reestablish the faint pink color. Read the burets to the nearest 0.01 mL, and record these final readings in the **Data Table.**

10. Discard the liquid in the flask, rinse the flask thoroughly with deionized water, and run a second and third trial.

11. Record the known concentration of the standard HCl solution in the **Data Table.**

DISPOSAL

12. Clean all apparatus and your lab station. Return equipment to its proper place. Dispose of chemicals and solutions in the containers designated by your teacher. Do not pour any chemicals down the drain or in the trash unless your teacher directs you to do so. Wash your hands thoroughly before you leave the lab and after all work is finished.

Data Table				
Buret readings (ml)				
	HCl		**NaOH**	
Trial	**Initial**	**Final**	**Initial**	**Final**
1				
2				
3				

Molarity of HCl _____

Analysis

1. Organizing Data Calculate the volumes of acid used in the three trials. Show your calculations and record your results below.

Trial 1: Volume of HCl = _____

Trial 2: Volume of HCl = _____

Trial 3: Volume of HCl = _____

Titration with an Acid and a Base *continued*

2. **Organizing Data** Calculate the volumes of base used in the three trials. Show your calculations and record your results below.

   Trial 1: Volume of NaOH = _____

   Trial 2: Volume of NaOH = _____

   Trial 3: Volume of NaOH = _____

3. **Organizing Data** Use equation **3** in the introduction to determine the number of moles of acid used in each of the three trials. Show your calculations and record your results below.

   Trial 1: Moles of acid = _____

   Trial 2: Moles of acid = _____

   Trial 3: Moles of acid = _____

4. **Relating Ideas** Write the balanced equation for the reaction between HCl and NaOH.

   _____

5. **Organizing Ideas** Use the mole ratio in the balanced equation and the moles of acid from **Analysis** item **3** to determine the number of moles of base neutralized in each trial. Show your calculations and record your results below.

   Trial 1: Moles of acid = _____

   Trial 2: Moles of acid = _____

   Trial 3: Moles of acid = _____

6. **Organizing Data** Use equation **2** in the introduction and the results of **Analysis** item **2** and **5** to calculate the molarity of the base for each trial. Show your calculations and record your results below.

   Trial 1: Molarity of NaOH _____

   Trial 2: Molarity of NaOH _____

   Trial 3: Molarity of NaOH _____

7. **Organizing Conclusions** Calculate the average molarity of the base. Show your calculations and record your result below.

| Titration with an Acid and a Base *continued*

Conclusions

1. Analyzing Methods In step **6,** you added deionized water to the HCl solution in the Erlenmeyer flask before titrating. Why did the addition of the water not affect the results?

_____

_____

_____

_____

2. Analyzing Methods What characteristic of phenolphthalein made it appropriate to use in this titration? Could you have done the experiment without it? How does phenolphthalein's end point relate to the equivalence point of the reaction?

_____

_____

_____

_____

_____

_____

_____

Hydronium Ion Concentration and pH

For pure water at 25°C, the hydronium ion concentration, $[H_3O^+]$, is 1.0×10^{-7} M or 10^{-7} M.

In *acidic* solutions, the hydronium ion concentration is *greater* than 1×10^{-7} M. For example, the H_3O^+ concentration of a 0.00 001 M hydrochloric acid (HCl) solution is 1×10^{-5} M.

In *basic* solutions, the hydronium ion concentration is *less* than 1×10^{-7} M. The hydronium ion concentration in a basic solution can be determined from the equation for K_w when the hydroxide ion concentration, $[OH^-]$, is known.

(1) $\qquad K_w = [H_3O^+][OH^-] = 1 \times 10^{-14}$ mol^2/L^2

(2) $\qquad [H_3O^+] = \dfrac{1 \times 10^{-14} \text{ mol}^2/\text{L}^2}{[OH^-]}$

Therefore, the concentration of H_3O^+ in a 0.000 01 M NaOH basic solution is

$$[H_3O^+] = \frac{1 \times 10^{-14} \text{ mol}^2/\text{L}^2}{[OH^-]} = \frac{1 \times 10^{-14} \text{ mol}^2/\text{L}^2}{0.000 \, 01 \text{ mol/L}} = \frac{1 \times 10^{-14} \text{ mol}^2/\text{L}^2}{1 \times 10^{-5} \text{ mol/L}}$$

$$= 1 \times 10^{-9} \text{ mol/L}$$

The pH of a solution is defined as the negative of the common logarithm of the hydronium ion concentration. Therefore, for a 1×10^{-5} M HCl solution,

(3) $\qquad$ pH $= -\log [H_3O^+]$

$\qquad\qquad\quad = -\log (1 \times 10^{-5}) = 5$

For a 1×10^{-5} M NaOH solution, from equation **2**,

$$[H_3O^+] = \frac{1 \times 10^{-14} \text{ mol}^2/\text{L}^2}{1 \times 10^{-5} \text{ mol/L}} = 1 \times 10^{-9} \text{ mol/L}$$

$$\text{and pH} = -\log (1 \times 10^{-9}) = 9$$

Values for pH that are greater than 7 indicate a basic solution. The higher the pH above the value of 7, the stronger the base and the smaller the hydronium ion concentration.

OBJECTIVES

Use pH paper and standard colors to determine the pH of a solution.

Determine hydronium ion concentrations from experimental data.

Describe the effect of dilution on the pH of acids and sodium hydroxide.

Relate pH to the acidity and basicity of solutions.

| Hydronium Ion Concentration and pH *continued*

MATERIALS

- 0.033 M H_3PO_4
- 0.10 M CH_3COOH
- 0.10 M HCl
- 0.10 M NaCl
- 0.10 M Na_2CO_3
- 0.10 M $NaHCO_3$
- 0.10 M NaOH
- 0.10 M NH_3, aqueous
- 0.10 M CH_3COONH_4

- 10 mL graduated cylinder
- 50 mL graduated cylinder
- 250 mL beaker
- deionized water
- glass plate
- glass stirring rod, 25 cm long
- pH papers, wide and narrow range
- test tubes 9
- white paper, small sheet
- test tube rack

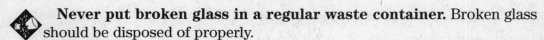

Always wear safety goggles and a lab apron to protect your eyes and clothing. If you get a chemical in your eyes, immediately flush the chemical out at the eyewash station while calling to your teacher. Know the locations of the emergency lab shower and the eyewash station and the procedures for using them.

Do not touch any chemicals. If you get a chemical on your skin or clothing, wash the chemical off at the sink while calling to your teacher. Make sure you carefully read the labels and follow the precautions on all containers of chemicals that you use. If there are no precautions stated on the label, ask your teacher what precautions you should follow. Do not taste any chemicals or items used in the laboratory. Never return leftovers to their original containers; take only small amounts to avoid wasting supplies.

Call your teacher in the event of a spill. Spills should be cleaned up promptly, according to your teacher's directions.

Never put broken glass in a regular waste container. Broken glass should be disposed of properly.

Hydronium Ion Concentration and pH *continued*

Procedure

1. Place the glass plate on the sheet of white paper. Place a strip of wide-range pH paper and a strip of narrow-range pH paper on the glass plate.

2. Obtain samples of all of the solutions listed in **Data Table 1**. To find the pH of each solution, dip a clean stirring rod into each solution and apply a drop of the solution first to the wide-range pH paper and then to the narrow-range pH paper. **Figure A** shows the correct technique. Compare the color produced by each solution with the colors on the charts included with the pH papers. Be sure to rinse and dry the stirring rod before you test each solution. Record your results in **Data Table 1**.

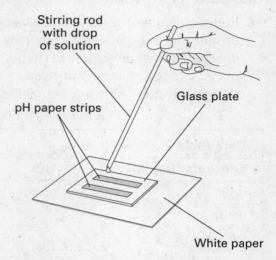

Stirring rod with drop of solution

Glass plate

pH paper strips

White paper

Figure A

3. Using the 10 mL graduated cylinder, measure 5.0 mL of 0.10 M HCl. Dilute it to 50 mL in the 50 mL graduated cylinder with deionized water added from the 10 mL graduated cylinder. Stir the solution with a stirring rod. Transfer 5.0 mL of the diluted solution to a labeled test tube. Using the 10 mL graduated cylinder, measure and save 5.0 mL of the diluted solution from the 50 mL graduated cylinder. Empty the 50 mL graduated cylinder into a waste container, and rinse it with deionized water.

4. Repeat step **3** two more times, starting with the 5.0 mL of *diluted* solution you just made and measured in the 10 mL graduated cylinder.

Data Table 1			
0.1 M solution	**pH**	**0.1 M solution**	**pH**
HCl		NaCl	
CH_3COOH		Na_2CO_3	
H_3PO_4		$NaHCO_3$	
NaOH		CH_3COONH_4	
NH_3			

Name _____ Class _____ Date _____

| Hydronium Ion Concentration and pH *continued*

5. Repeat steps **3** and **4** for 0.10 M NaOH and 0.10 M CH₃COOH. Record your results for each test in **Data Table 2.**

DISPOSAL

6. Clean all apparatus and your lab station. Return equipment to its proper place. Dispose of chemicals and solutions in the containers designated by your teacher. Do not pour any chemicals down the drain or in the trash unless your teacher directs you to do so. Wash your hands thoroughly after all work is finished and before you leave the lab.

Data Table 2

Concentration (M)	HCl	NaOH	CH₃COOH
0.10			
0.010			
0.0010			
0.000 10			

Analysis

1. Organizing Data List the solutions in order of decreasing acid strength using your results from step **2.**

_____

_____

_____

_____

2. Organizing Data Calculate the theoretical pH values for the concentrations prepared in steps **3–5.** Record these values below.

HCl	Calculated pH	NaOH	Calculated pH
0.10 M		0.10 M	
0.010 M		0.010 M	
0.0010 M		0.0010 M	
0.000 10 M		0.000 10 M	

Hydronium Ion Concentration and pH *continued*

3. Analyzing Results What effect does dilution have on the pH of (a) an acid and (b) a base?

a. _____

_____

b. _____

_____

Conclusions

1. Predicting Outcomes Solutions with a pH of 12 or greater dissolve hair. Would a cotton shirt or a wool shirt be affected by a spill of 0.1 M sodium hydroxide? Explain.

_____

_____

_____

_____

_____

_____

_____

Temperature of a Bunsen Burner Flame

When a hot solid is immersed in a cool liquid, heat flows from the hot object to the cool liquid. In fact, the number of joules of energy lost by the hot solid (ΔQ_1) equals the number of joules of energy gained by the cool liquid (ΔQ_2).

$$-\Delta Q_1 = +\Delta Q_2$$

The quantity of energy that is lost or gained is a function of the object's mass and specific heat. Every material has a characteristic specific heat, C_p, which is the number of joules of energy needed to change the temperature of 1.0 g of material 1.0°C. Some specific heats are given in **Table 1.** If the mass, temperature change, and specific heat of a substance are known, the energy lost or gained can be calculated using this relationship: $\Delta Q = (mass)(specific\ heat)(\Delta T)$. The temperature change is defined as $\Delta T = T_{final} - T_{initial}$.

In this experiment, you will determine the temperature of a Bunsen burner flame by heating a sample of a known metal in the flame and then immersing the hot metal at temperature T_1 into a measured quantity of water at temperature T_2. As the energy flows from the hot metal to the cool water, the two materials approach an intermediate temperature T_3. These changes are shown graphically in **Figure A.**

$$-\Delta Q_1 = \Delta Q_2$$

$$-(m_1)(c_{p1})(T_3 - T_1) = (m_2)(c_{p2})(T_3 - T_2)$$

By solving the equation for T_1, you can find the initial temperature of the hot metal. The metal should be the same temperature as the Bunsen burner flame.

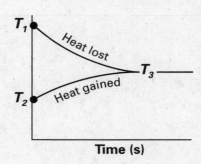

Figure A

OBJECTIVES

Observe the change in temperature of a known mass of water when a heated metal object of known mass is placed in it.

Use the specific heat of the metal and of water to calculate the initial temperature of the object.

Relate the temperature of the object to the temperature of the Bunsen burner flame.

| Temperature of a Bunsen Burner Flame *continued*

MATERIALS

- balance
- Bunsen burner
- foam cup
- iron ring

- metal object (Fe or Cu), 20–30 g
- Nichrome wire, 20 cm
- ring stand
- thermometer, nonmercury

 Always wear safety goggles and a lab apron to provide protection for your eyes and clothing. If you get a chemical in your eyes, immediately flush the chemical out at the eyewash station while calling to your teacher. Know the locations of the emergency lab shower and the eyewash station and the procedures for using them.

When using a Bunsen burner, confine long hair and loose clothing. If your clothing catches on fire, **walk** to the emergency lab shower and use it to put out the fire. Do not heat glassware that is broken, chipped, or cracked. Use tongs or a hot mitt to handle heated glassware and other equipment because hot glassware does not look hot.

Never put broken glass in a regular waste container. Broken glass should be disposed of properly.

Procedure

CAUTION: Remember that the metal object will become very hot.

1. Obtain the mass of an empty foam cup to the nearest 0.01 g, and record the mass in the **Data Table.**

2. Fill the cup about two-thirds full with water. Obtain the mass of the cup and water and record it in the **Data Table.**

3. Read and record the temperature of the water to the nearest 0.1°C.

4. Obtain the mass of the metal object to the nearest 0.01 g and record it in the **Data Table.**

5. Use the Nichrome wire to attach the metal object to the iron ring clamped on the ring stand. The object should hang about 7 or 8 cm below the ring.

6. Adjust the Bunsen burner so that it produces a hot blue flame. Move the flame under the metal object, and heat it for about 5 min.

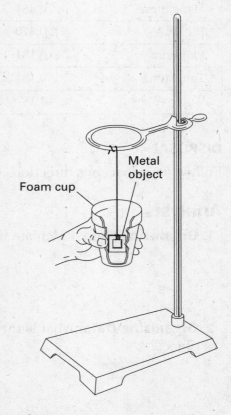

Figure A

| Temperature of a Bunsen Burner Flame *continued*

7. Turn off the Bunsen burner and move it aside. SLOWLY lift the foam cup containing the water so that the hot metal becomes immersed in the water, as shown in **Figure B.** Hold the cup in this position for 2 min, stir the water, and then read and record the temperature of the water in the **Data Table.**

Data Table	
Type of metal used	
Mass of metal	g
Specific heat of metal (from **Table 1**)	J/(g·°C)
Mass of empty foam cup	g
Mass of foam cup and water	g
Initial temperature of water	°C
Final temperature of water	°C

Table 1: Specific Heats of Materials	
Substance	Specific heat, J/(g·°C)
Copper	0.385
Iron	0.444
Manganese	0.481
Nickel	0.470
Platinum	0.131
Tungsten	0.134
Water	4.18

DISPOSAL

Follow your teacher's directions.

Analysis

1. Organizing Data Calculate the change in the temperature of the water.

2. Organizing Data What is the mass of the water that was heated by the metal object?

Temperature of a Bunsen Burner Flame *continued*

3. Organizing Data Calculate the energy gained by the water. (Hint: Refer to the introduction.)

4. Inferring Conclusions Calculate the temperature of the hot metal object. (Hint: Refer to the introduction. The energy lost by the metal equals the energy gained by the water.)

5. Inferring Conclusions What is the temperature of the Bunsen burner flame?

_____

6. Analyzing Information From the specific heats of metals listed in **Table 1,** which metal would raise the temperature of the water the most in this experiment? the least?

_____

_____

7. Evaluating Methods Error is present in every experiment. In this experiment, the small mass of water that was changed to steam when the hot sample was initially immersed in the water was neglected. Would this tend to make the calculated temperature lower or higher than the actual temperature of the flame. Explain your answer.

_____

_____

_____

8. Designing Experiments It takes about 2.26 kJ of energy to evaporate each gram of water. Suggest a way to include the evaporated water in your calculation. (Assuming no water is ejected when the metal is inserted into the water.)

_____

_____

_____

_____

_____

Temperature of a Bunsen Burner Flame *continued*

9. Evaluating Methods Another source of error is the small amount of hot Nichrome wire that was heated and immersed in the water along with the metal object. Would this tend to make the calculated temperature lower or higher than the actual temperature of the flame? Explain your answer.

_____

_____

_____

Conclusions

1. Predicting Outcomes If 100 g of boiling water at 100°C is poured into a 100 g platinum beaker at 0°C, an intermediate temperature will result. Knowing that the specific heat of water is about 32 times greater than the specific heat of platinum, approximate the intermediate temperature.

_____

_____

_____

2. Applying Ideas In the eighteenth century, clothes were pressed using the heat from a heavy piece of metal with a flat, smooth side. This solid metal "iron" had to be heated periodically on top of a wood- or coal-burning stove. Disregarding cost, explain which of the metals listed in **Table 1** would be the best to use for this purpose.

_____

_____

_____

Name _____ Class _____ Date _____

Counting Calories

We expend energy in three ways—through exercise, through specific dynamic action (SDA), and through basal metabolism (BM). Exercise is the composite of all physical work we do with our bodies. SDA accounts for the energy consumed in digesting and metabolizing food. BM accounts for the energy needed to maintain key bodily functions that sustain life.

Food provides energy. Food chemicals are classified as fats and oils, carbohydrates, or proteins. Fats furnish 9 Calories (Cal) per gram, carbohydrates furnish 4 Cal per gram, and proteins 4 Cal per gram. The amount of energy each of these chemicals delivers to the body is independent of the kind of food or quantity of other macronutrients. A meal containing 10 g of fat, 15 g of carbohydrates, and 20 g of protein would deliver 230 Cal of energy: 90 Cal from fat, 60 Cal from carbohydrates, and 80 Cal from protein. In this experiment, you will construct a calorimeter and indirectly determine the caloric content of peanuts and walnuts by measuring the temperature change in a sample of water.

Note: The *C* in nutritional *Calorie* is capitalized. The nutritional Calorie is equal to 1000 calories, or 1 kilocalorie.

OBJECTIVES

Construct a calorimeter setup.

Determine the amount of energy, in Calories, in oil-roasted walnuts and peanuts.

Compare measured caloric data.

Determine which nut is a better source of energy.

MATERIALS

- peanut halves
- walnut halves
- heavy aluminum foil, 5 × 5 cm
- 250 mL beaker
- 100 mL graduated cylinder
- aluminum pie plate
- can opener
- balance
- cork stopper
- digital temperature recorder or nonmercury thermometer
- knife or single-edge razor blade
- butane lighter
- metal file
- metric ruler
- paper clip
- pot holder
- 16 oz tin can, opened and clean
- tin snips

Always wear safety goggles and a lab apron to provide protection for your eyes and clothing. If you get a chemical in your eyes, immediately flush the chemical out at the eyewash station while calling to your teacher. Know the locations of the emergency lab shower and the eyewash station and the procedures for using them.

Do not touch any chemicals. If you get a chemical on your skin or clothing, wash the chemical off at the sink while calling to your teacher. Make
sure you carefully read the labels and follow the precautions on all containers of
chemicals that you use. If there are no precautions stated on the label, ask your
teacher what precautions you should follow. Do not taste any chemicals or items
used in the laboratory. Never return leftovers to their original containers; take only
small amounts to avoid wasting supplies.

Call your teacher in the event of a spill. Spills should be cleaned up
promptly, according to your teacher's directions.

Never put broken glass in a regular waste container. Broken glass
should be disposed of properly.

Never stir with a thermometer, because the glass around the bulb is fragile and might break.

When using a flame, confine long hair and loose clothing. If your
clothing catches on fire, WALK to the emergency lab shower, and use it to
put out the fire. Do not heat glassware that is broken, chipped, or cracked. Use a
hot mitt or pot holder to handle heated glassware and other equipment, because
hot glassware does not look hot.

Procedure

Before beginning the lab, inform your teacher if you have an allergy to peanuts or
walnuts.

1. Use **Figure A** to prepare the retort. Remove the label from the tin can. Use
   the can opener to punch four triangular openings in the unopened end (top)
   of the can and three openings in the side of the can, near the top. Use tin
   snips to cut a viewing hole in the side of the can, beginning at the can's open
   end: 5 cm high, 3 cm wide at the top, and 4 cm wide at the bottom. Use a
   metal file to remove all burrs and sharp edges.

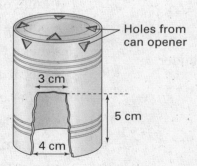

Holes from
can opener

3 cm

5 cm

4 cm

Figure A

Counting Calories *continued*

2. Refer to **Figure B** to construct the sample holder. Bend a paper clip so that it can hold food samples between the wires. Measure and cut the cork stopper so that its height, when resting on its widest flat surface, is 2 cm. Mold a piece of aluminum foil around the cork. Insert the bent paper clip into the cork. The total height of the holder assembly should be no higher than 3.5 to 4.0 cm.

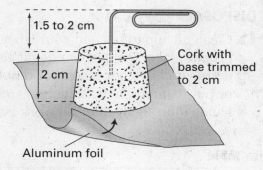

1.5 to 2 cm

2 cm

Cork with base trimmed to 2 cm

Aluminum foil

Figure B

3. Obtain three peanut halves and three walnut halves to use as your six samples. Use a balance to determine the mass of each sample to the nearest 0.1 g. For each sample, record the type of nut and its mass in the **Data Table.**

4. Using a graduated cylinder, pour 100 mL of tap water into the 250 mL beaker.

5. Insert the temperature recorder into the beaker. Make sure that the sensing element is fully submerged and does not make contact with the bottom glass. If a digital temperature recorder is unavailable, use a nonmercury thermometer.

6. Measure the temperature of the water in the beaker; record this value in the **Data Table.**

7. Place a sample in the wire of your sample holder. Position the sample holder in the center of the aluminum pie plate.

8. Using a butane lighter, carefully set fire to the nut. Once burning is sustained, carefully position the retort (metal can with holes) over the burning sample so that the viewing hole faces you. Set the water-filled beaker on top of the retort. Refer to **Figure C.**

9. Carefully stir the water and observe the digital readout of the temperature until it starts to fall. Record the maximum temperature value in the **Data Table.**

10. Cool the materials for 2 to 5 min. Using a pot holder, remove the beaker from the retort and pour the water into the sink. Using a pot holder, remove the retort from the pie plate.

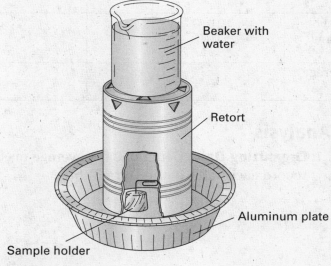

Beaker with water

Retort

Aluminum plate

Sample holder

Figure C

11. Repeat steps **4** through **9** until you have collected data for three peanut halves and three walnut halves.

DISPOSAL

12. Clean all apparatus and your lab station. Return equipment to its proper place. Dispose of chemicals and solutions in the containers designated by your teacher. Do not pour any chemicals down the drain or put them in the trash unless your teacher directs you to do so. Wash your hands thoroughly after all work is finished and before you leave the lab.

Data Table

	Mass of sample (g)	Before burning (T₁)	After burning (T₂)	Change in water temperature	calories	kcal	kcal per 100 g of food sample
		Water temperature (°C)			**Food energy**		
Trial 1 peanut							
Trial 2 peanut							
Trial 3 peanut							
Average							
Trial 1 walnut							
Trial 2 walnut							
Trial 3 walnut							
Average							

Analysis

1. Organizing Data Determine the average mass for each nut type tested; record each value in the **Data Table**.

2. **Organizing Data** Calculate the change in water temperature for each trial in the **Data Table** by subtracting T_1 from T_2. Enter the results in the **Data Table.** Then calculate the average change in water temperature for each nut type tested; record the values in the **Data Table.**

3. **Organizing Data** Determine the number of Calories (kcal) contained in 100 g of food sample for both nut types tested; record the values in the **Data Table.** The energy transferred by the combustion of the nut equals the energy absorbed by the water. Use 1 cal/(g·°C) as the specific heat capacity of water when you calculate the number of calories.

4. **Predicting Outcomes** Suppose a package of lunch meat contains 5% fat according to the label. This means that 5% of the mass is fat. What percentage of the total calories does this amount of fat provide? Assume that the lunch meat has a mass of 100 g and is 5% fat by mass.

Counting Calories *continued*

Conclusion

1. Inferring Conclusions According to your experimental data, which of the nuts that you tested is a better source of energy? Why?

_____

_____

_____

Radioactivity

aThe types of radiation emitted by radioactive materials include alpha particles, beta particles, and gamma rays. Alpha particles are helium nuclei; beta particles are high-speed electrons; and gamma rays are extremely high-frequency photons. Alpha particles will discharge an electroscope, but their penetrating power is not great enough to affect a Geiger counter tube. A few centimeters of air will stop an alpha particle. Beta particles can be detected by a Geiger counter tube and can penetrate several centimeters of air, but several layers of paper or aluminum foil can stop them. Gamma rays can penetrate through several centimeters of concrete.

The environment contains a small amount of natural radiation, which can be detected by a Geiger counter. This is called background radiation and is primarily due to cosmic rays from stars. A small amount of background radiation may also come from the walls of buildings made of stone, clay, and some kinds of bricks, and from other sources such as dust particles.

OBJECTIVES

Use a Geiger counter to determine the level of background radiation.

Compare counts per minute for a beta source passing through air, index cards, and aluminum.

Graph data and determine the relationship between counts per minute and thickness of material.

Graph data and determine the effect of distance on counts per minute in air.

MATERIALS

- aluminum foil
- index cards
- radioactivity demonstrator (scaler)
- thallium 204 (beta source)

Always wear safety goggles and a lab apron to protect your eyes and clothing. If you get a chemical in your eyes, immediately flush the chemical out at the eyewash station while calling to your teacher. Know the location of the emergency lab shower and eyewash station and the procedures for using them.

Do not touch any chemicals. If you get a chemical on your skin or clothing, wash the chemical off at the sink while calling to your teacher. Make sure you carefully read the labels and follow the precautions on all containers of chemicals that you use. If there are no precautions stated on the label, ask your teacher what precautions to follow. Do not taste any chemicals or items used in the laboratory. Never return leftovers to their original container; take only small amounts to avoid wasting supplies.

Name _____ Class _____ Date _____

Radioactivity *continued*

Procedure
PART–BACKGROUND COUNT

CAUTION The wiring between the Geiger counter probe and the counter carries more than 1000 volts. Do not touch it when the equipment is operating or for at least 5 min after it has been unplugged from the electric socket.

1. Put on safety goggles, gloves, and lab apron.

2. Carefully read the directions on the operation of your counter. Set the counter to zero. Do not have any radioactive sources within 1 m of the Geiger counter tube. Turn on the counter for 1 min and count the frequency of clicks. Record the number of clicks per minute in **Table 1.** Repeat two more trials. The average of these trials will be your background count in counts per minute (cpm).

PART–RANGE OF BETA PARTICLES IN VARIOUS MEDIA

3. Air: Place the beta source approximately 5 cm from the Geiger tube. The proper setup is shown in **Figure 1.**
   CAUTION You should never directly handle a radioactive source. Be certain you are wearing gloves when handling the beta source (thallium 204).

   Determine the count for 1 min, and record your count/per minute and distance between the source and Geiger tube in **Table 2.**

4. Move the source another 5 cm, and determine and record the count for 1 min. Continue this procedure until you reach the background count or have completed 10 trials.

Figure 1

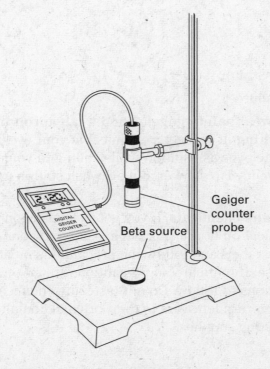

Geiger counter probe

Beta source

▌Radioactivity *continued*

5. Paper: Place the beta source approximately 5 cm from the Geiger tube. Place a single index card on top of the beta source and determine the count for 1 min. Record the number of counts per minute in **Table 3.**

6. Keep the distance from the source to the Geiger tube constant, and repeat step **4** with an additional index card. Continue adding index cards and measuring and recording counts until the background count is reached or 10 trials have been completed.

7. Aluminum: Repeat steps **4** and **5,** but use sheets of aluminum foil in place of index cards. Record your data in **Table 3.**

8. Return the beta source to your teacher.

TABLE 1 BACKGROUND COUNT DATA

Background Count	
Trial	Count/min
1	
2	
3	
Average	

TABLE 2 RANGE OF BETA PARTICLES IN AIR

Air					
Distance (cm)	Count/ min	Distance (cm)	Count/ min	Distance (cm)	Count/ min

TABLE 3 PENETRATING POWER OF BETA PARTICLES

Index Cards		Aluminum	
No. Sheets	**Count/min**	**No. Sheets**	**Count/min**
0		0	
1		1	
2		2	
3		3	
4		4	
5		5	
6		6	
7		7	
8		8	
9		9	
10		10	

Analysis

1. **Organizing Data** On a separate piece of graph paper, plot a graph of counts per minute versus distance between the beta source and the Geiger tube. Subtract the background count from each of the readings. Place counts per minute on the vertical axis and the distance on the horizontal axis. Describe the relationship between distance and counts per minute.

_____

_____

2. **Organizing Data** On a separate piece of graph paper, plot a graph of the number of index cards versus counts per minute. Place the counts per minute on the vertical axis. Describe the relationship between the number of index cards and counts per minute.

_____

_____

3. **Organizing Data** On a separate piece of graph paper, plot a graph of the number of sheets of aluminum foil versus counts per minute. Place the counts per minute on the vertical axis. Describe the relationship between the number of sheets of aluminum and counts per minute.

_____

_____

| Radioactivity *continued*

Conclusions

1. **Analyzing Results** According to the three graphs, which substance is the most efficient absorber of beta particles? Explain how you made your decision.

_____

_____

2. **Relating Ideas** What is the implication of discovering a larger-than-normal amount of helium in a natural-gas well?

_____

_____

_____

_____

Skills Practice

Detecting Radioactivity

The element radon is the product of the radioactive decay of uranium. The $^{222}_{86}Rn$ nucleus is unstable and has a half-life of about 4 days. Radon decays by giving off alpha particles (helium nuclei) and beta particles (electrons) according to the equations below, with $^{4}_{2}He$ indicating an alpha particle and $^{0}_{-1}\beta$ representing a beta particle. Chemically, radon is a noble gas. Other noble gases can be inhaled without causing damage to the lungs because these gases are not radioactive. When radon is inhaled, however, it can rapidly decay into polonium, lead, and bismuth, all of which are solids that can lodge in body tissues and continue to decay. The lead isotope shown at the end of the chain of reactions, $^{210}_{82}Pb$, has a half-life of 22.6 years, and it will eventually undergo even more decay before creating the final stable product, $^{206}_{82}Pb$.

$$^{222}_{86}Rn$$
$$\searrow$$
$$^{218}_{84}Po + {}^{4}_{2}He$$
$$\searrow$$
$$^{214}_{82}Pb + {}^{4}_{2}He$$
$$\searrow$$
$$^{214}_{83}Bi + {}^{0}_{-1}e$$
$$\searrow$$
$$^{214}_{84}Po + {}^{0}_{-1}e$$
$$\searrow$$
$$^{210}_{82}Pb + {}^{4}_{2}He \rightarrow$$

In this experiment, you will measure the level of radon emissions in your community. First you will construct a simple detector using plastic (CR-39) that is sensitive to alpha particles. You will place the detector in your home or somewhere in your community for a 3 week period. Then, in the lab, the plastic will be etched with sodium hydroxide to make the tracks of the alpha particles visible. You will examine the tracks to determine the number of tracks per cm^2 per day and the activity of radon at the location. Finally, all class data will be pooled to make a map showing radon activity throughout your community.

OBJECTIVES

Build a radon detector and use it to detect radon emissions.

Observe the tracks of alpha particles microscopically and count them.

Calculate the activity of radon.

Evaluate radon activity over a large area using class data and draw a map of its activity.

Detecting Radioactivity *continued*

MATERIALS

- clear plastic ruler or stage micrometer
- CR-39 plastic
- etch clamp (the ring from a key chain)
- index card, 3 in. × 5 in.
- microscope
- paper clips
- push pin
- scissors

- small plastic cup with lid
- tape
- toilet paper or other tissue

Optional

- sources of radiation (Fiestaware, Coleman green-label lantern mantles, old glow-in-the-dark clock or watch faces, cloud-chamber needles)

 Always wear safety goggles and a lab apron to protect your eyes and clothing. If you get a chemical in your eyes, immediately flush the chemical out at the eyewash station while calling to your teacher. Know the location of the emergency lab shower and eyewash station and the procedures for using them.

Scissors and push pins are sharp; use with care.

Procedure

PART 1–DETECTOR CONSTRUCTION

1. Put on safety goggles, gloves, and lab apron.

2. Cut a rectangle, 2 cm × 4 cm, from the index card.

3. Locate the side of the CR-39 plastic that has the felt-tip marker lines on it. Peel off the polyethylene film, and use the push pin to inscribe a number or other identification near the edge of the piece. With a short piece of transparent tape, form a loop with the sticky side out. Place the tape on the back of the piece of CR-39 plastic (the side that is still covered with polyethylene), and firmly attach it to the index-card rectangle.

4. With a permanent marker, write the ID number or other identification on the outside of the plastic cup. Place the paper rectangle, with the CR-39 plastic on top, into the cup.

5. Cut a hole in the center of the plastic-cup lid. Place a small piece of tissue over the top of the cup to serve as a dust filter. Then snap the lid onto the cup.

6. Place the completed detector in the location of your choice. (Check with your teacher first.) Record this information in **Table 1.** The detector must remain undisturbed at that location for at least 3 weeks.

7. At the end of the 3 weeks, return the entire detector to your teacher for the chemical-etching process and the counting of the radiation tracks.

| Detecting Radioactivity *continued*

PART 2–ETCHING

8. Remove the CR-39 plastic from the plastic cup, detach it from the index card, and peel the polyethylene film from the back. Slip the ring of an etch clamp over the top of the CR-39 plastic, and hook it onto a large paper clip that has been reshaped to have "hooks" at each end, as shown in **Figure 1.**

9. When you have completed this work, give the plastic to your teacher for the etching step. During this step, NaOH solution will be used to remove the outer layer of the plastic so that the tracks of the alpha particles become visible.

PART 3–COUNTING THE TRACKS

10. Examine **Figure 2** below. Notice the various shapes of the tracks left as alpha particles entered the CR-39 plastic. The circular tracks were formed by alpha particles that entered straight on, and the teardrop-shaped tracks were formed by alpha particles that entered at an angle.

11. Place your plastic sample under a microscope to view the tracks. Use a clear-plastic metric ruler or a stage micrometer to measure the diameter of the microscope's field of view. Make this measurement for low power (10×). Record this diameter in **Table 1.**

12. The tracks are on the top surface of the CR-39 plastic. Make certain that you focus on that surface and the tracks look like those in the illustration. These tracks were produced by placing the CR-39 plastic within a radium-coated clay urn. Your radon detector should not have nearly as many tracks. If there are too many tracks, switch to a high-power (40×) objective, and measure and record the diameter of the field of view. Place your piece of CR-39 plastic on a microscope slide. Count and record the number of tracks in 10 different fields. Record these numbers in **Table 2.**

Figure 1

Figure 2

Alpha particle tracks

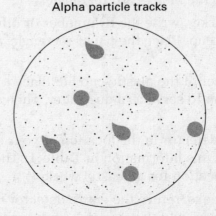

| Detecting Radioactivity *continued*

TABLE 1 DETECTOR TIME AND LOCATION DATA

Date detector was put in place	
Time detector was put in place	
Location of detector (e.g., bedroom, kitchen, dining room, furnace room, under the stairwell, etc.)	
Floor level of location	
Date detector was removed	
Time detector was removed	
Diameter of microscope field	

TABLE 2 ALPHA PARTICLE TRACK DATA

Field										
Number of tracks										

Analysis

1. **Organizing Data** In order to increase the accuracy of your data, you counted tracks in 10 different areas. Find the average number of tracks in a single area for your piece of CR-39 plastic.

2. **Organizing Data** You counted the number of tracks within the field of view of your microscope. In order to calculate the number of tracks per cm^2, you need to know the area of the field. Use the diameter of the field to calculate the area. (Hint: The field is circular; $d = 2r$; $A = \pi r^2$.)

3. **Organizing Data** Using the answers to Analysis items **1** and **2**, calculate the average number of tracks per cm^2.

| Detecting Radioactivity *continued*

4. Organizing Data In Analysis item **3** you calculated the number of tracks that accumulated per cm^2 over the total period the detector was in place. Divide this number by the number of days the detector was in place to calculate tracks/cm^2/day.

5. Organizing Data Several pieces of CR-39 plastic were sent to a facility that had a radon chamber in which the activity of the radon was known to be 13.69 Bq/L (becquerels per liter of air), as measured by a different technique. A becquerel is the name for the SI unit of activity for a radioactive substance and is equal to 1 decay/s. The exposed pieces of CR-39 plastic were etched, counted, and found to have an activity of 2370 tracks/cm^2/day. Calculate the radon activity measured by your detector in becquerels/liter. (Hint: Use the proportion 13.69 Bq/L:2370 tracks/cm^2/day as a conversion factor to convert your data.)

6. Relating Ideas The Environmental Protection Agency and other U.S. government agencies use non-SI units of picocuries per liter (pCi/L) to measure radiation. One curie is 3.7×10^{10} Bq, and one picocurie is 10^{-12} curies. Convert your data into units of pCi/L.

7. Organizing Data Combine your data with those of other teams in your class, and jointly construct a map of your region that shows the levels of radon activity in Bq/L and pCi/L at various locations throughout your community.

| Detecting Radioactivity *continued*

Conclusions

1. Applying Ideas The half-life of $^{222}_{86}Rn$ is 3.823 days. After what time will only one-fourth of a given amount of radon remain?

_____

_____

_____

_____

2. Designing Experiments Factors other than geographic location can have an effect on radon emissions. For example, readings taken in basements are likely to be higher than those in attics. Design an experiment to test different aspects of the three highest and three lowest regions of radiation on the map to examine the influence of these other factors. If your teacher approves your suggestion, try it.

_____

_____

_____

_____

_____

3. Designing Experiments CR-39 plastic could be used for further investigations of naturally occurring radiation. Design an experiment to explore one of the following areas. If your teacher approves your plan, carry out the experiment.

• the range of alpha particles

• the radon activity in soil

• radioactivity of common items such as lantern mantles (Coleman green label), glow-in-the-dark clock or watch faces, and pieces of Fiestaware

_____

_____

_____

_____

_____

Name _____ Class _____ Date _____

Skills Practice

Carbon

Carbon exhibits allotropy, the existence of an element in two or more forms in the same physical state. Diamond and graphite are carbon's two crystalline allotropic forms. They result from the different ways in which carbon atoms link to one another in the two forms. A diamond contains an enormous number of carbon atoms that form an extremely strong, tetrahedral network. In graphite, the carbon atoms are bonded in a hexagonal pattern and lie in planes.

Microscopic graphite structures are contained in the black residues known as amorphous carbons that are obtained by heating certain carbon substances. Examples of amorphous carbon include charcoal, coke, bone black, and lamp-black. The amorphous forms are produced by a variety of procedures, including destructive distillation and incomplete combustion. If a carbon-rich compound such as gasoline, C_8H_{18}, is burned completely, the only products will be carbon dioxide and water vapor.

$$2C_8H_{18}(l) + 25O_2(g) \rightarrow 16CO_2(g) + 18H_2O(g)$$

Unfortunately, most internal combustion engines that run on gasoline are far from perfect. Incomplete combustion often results in black smoke (partially burned hydrocarbons) and carbon monoxide in the exhaust gases.

Carbon is an important reducing agent. The uses of its different varieties are related to particular properties, including combustibility and adsorption.

OBJECTIVES

Construct models of diamond and graphite.

Compare the observed physical properties of amorphous carbon with its molecular structure.

Infer why activated carbon is a good adsorbing agent.

100

MATERIALS

- activated charcoal, (for decolorizing)
- balance
- beaker, 250 mL
- Bunsen burner and related equipment
- buret clamp
- charcoal wood splinters
- crucible and cover
- dark-brown sugar solution
- Erlenmeyer flask, 125 mL
- filter paper
- forceps
- funnel
- funnel rack
- graduated cylinder, 50 mL

- HCl, 1.0 M
- iron ring
- limewater
- molecular model kits (2)
- NaOH, 1.0 M
- pipe-stem triangle
- ring stand
- rubber stopper, solid No. 4
- rubber stoppers, solid No. 2 (5)
- sugar, white granular
- test tube, 13 mm × 100 mm
- test tube, 25 mm × 100 mm (6)
- test-tube holder
- wooden splints

Always wear safety goggles, gloves, and a lab apron to protect your eyes and clothing. If you get a chemical in your eyes, immediately flush the chemical out at the eyewash station while calling to your teacher. Know the location of the emergency lab shower and eyewash station and the procedures for using them.

Do not touch any chemicals. If you get a chemical on your skin or clothing, wash the chemical off at the sink while calling to your teacher. Make sure you carefully read the labels and follow the precautions on all containers of chemicals that you use. If there are no precautions stated on the label, ask your teacher what precautions to follow. Do not taste any chemicals or items used in the laboratory. Never return leftovers to their original container; take only small amounts to avoid wasting supplies.

Call your teacher in the event of a spill. Spills should be cleaned up promptly, according to your teacher's directions.

Acids and bases are corrosive. If an acid or base spills onto your skin or clothing, wash the area immediately with running water. Call your teacher in the event of an acid spill. Acid or base spills should be cleaned up promptly.

Do not heat glassware that is broken, chipped, or cracked. Use tongs or a hot mitt to handle heated glassware and other equipment because hot glassware does not always look hot.

When using a Bunsen burner, confine long hair and loose clothing. If your clothing catches on fire, WALK to the emergency lab shower and use it to put out the fire.

When heating a substance in a test tube, the mouth of the test tube should point away from where you and others are standing. Watch the test tube at all times to prevent the contents from boiling over.

 Never put broken glass in a regular waste container. Broken glass should be disposed of separately according to your teacher's instructions.

Procedure

1. Put on safety goggles, gloves, and a lab apron.

2. Use a molecular model kit to start on the construction of the graphite model by constructing two hexagons using only the short connectors. Recall that the distances between the centers of adjacent carbon atoms in a layer of graphite are identical (142 pm). With **Figure 1** as a guide, connect the two hexagons using the longer stick connectors to represent the distance between centers in adjacent layers (335 pm). Use the remaining carbon spheres to build up each layer. Two carbons in the first hexagons you constructed will be common to the newly constructed hexagons. Use long connectors as needed between the rest of the constructed layers.

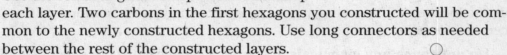

Graphite

Figure 1

3. Use 20 spheres representing carbon atoms from the molecular model kit to construct a model of diamond. Connect the spheres by means of the short stick connectors in the arrangement of atoms shown in **Figure 2.** How many carbon atoms is each carbon bonded to in the hexagon layer of graphite? What is the nature of the bonding between layers in graphite? How many carbon atoms is each carbon bonded to in diamond? Which structure is more rigid?

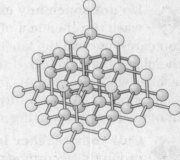

Diamond

Figure 2

Observations:

_____

_____

_____

_____

| **Carbon** *continued*

4. Ignite a wood splint in the burner flame. Note how it burns.

5. Break several splints and place the pieces in the bottom of a test tube. Clamp the test tube to the ring stand so that its mouth is slightly downward, as shown in **Figure 3.** Place a piece of paper on the table under the mouth of the test tube and heat the contents strongly until no more volatile matter is released.

Figure 3

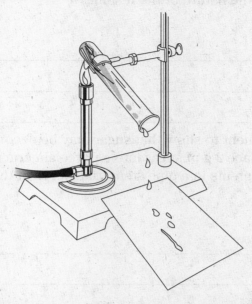

6. Remove the residue from the test tube carefully. Describe all the products of the reaction you observed.

Observations:

_____

_____

_____

7. Using forceps, ignite one piece of the solid residue in the burner and observe how it burns.

Observations:

_____

_____

_____

Carbon continued

8. Pour one inch of limewater into a small test tube. Using forceps, ignite another piece of solid residue, and insert the ignited end into the test tube, holding it above the limewater. Remove the ignited piece. Immediately stopper and shake the test tube. If nothing happens to the limewater, repeat with another piece of ignited residue. What happens to the limewater? What does this indicate about the nature of the residue?

Observations:

_____

_____

_____

9. Conduct an experiment to show that sugar may be decomposed into carbon and water vapor. Place 2 g of sugar into a porcelain crucible and cover it. Set the crucible and contents in a pipe-stem triangle. Heat the sugar on a low flame.

Observations:

_____

_____

_____

10. Using a test-tube holder, hold a test tube horizontally with its closed end in a yellow burner flame. What form of carbon is produced?

Observations:

_____

_____

_____

11. Change to the blue flame and hold the test tube so that the black deposit is in the hottest part of the flame.

Observations:

_____

_____

CAUTION Be careful when handling the acids and bases in step 12. Avoid contact with skin and eyes. If any of these chemicals should spill on you, immediately flush the area with water and then notify your teacher.

Carbon *continued*

12. Conduct experiments to determine the activity and solubility of wood charcoal splinters in dilute hydrochloric acid, water, and sodium hydroxide solution. Label each of three test tubes with the name of one of the test solutions and place 5 mL of the appropriate solution in the test tube. Place charcoal splints into each of the solutions. Stopper the test tubes and shake the contents. Describe your results. Dispose of the liquids and any pieces of charcoal as directed by your teacher.

Observations:

_____

_____

_____

_____

13. Pour 50 mL of dark-brown sugar solution into a 125 mL Erlenmeyer flask. Add 1 g of powdered activated charcoal and stopper the flask. Shake the mixture vigorously for 1 min.

14. Set up a funnel with filter paper, and filter the mixture. Pour the filtrate through the funnel a second time, and even a third time, if necessary, to remove the color from the sugar solution. Finally, compare the color of the filtrate with that of the original brown-sugar solution.

Observations:

_____

_____

15. Clean all apparatus and your lab station. Return equipment to its proper place. Dispose of chemicals and solutions in the containers designated by your teacher. Do not pour any chemicals down the drain or in the trash unless your teacher directs you to do so. Wash your hands thoroughly before you leave the lab and after all work is finished.

Analysis

1. Inferring Conclusions Of what use is activated charcoal in water purification processes?

_____

_____

2. Analyzing Information What is adsorption? Why are certain forms of charcoal good adsorbing agents?

_____

_____

_____

_____

_____

_____

_____

Conclusions

1. Analyzing Results In this experiment you used several different methods to prepare samples of carbon. What properties do all of the samples have in common? How does the model you built in **step 2** explain these properties?

_____

_____

2. Analyzing Ideas Automobile engineers use the ratio of carbon monoxide to carbon dioxide in exhaust gases as a measure of the efficiency of an engine. Which is better, a high ratio or a low ratio? Explain your answer. (Hint: Refer to the equation for the complete combustion of gasoline in the Introduction.)

_____

_____

_____

_____

Skills Practice

Polymers

Polymers are giant molecules consisting of repeating monomers, which are groups of atoms that form chains that are thousands of atoms long. Because these long molecular chains are linked by intermolecular forces, they can be molded into useful objects. If short bridges of atoms form between long polymeric chains, the polymer is then said to be cross-linked. Cross-linking gives the polymer new properties. In this experiment, you will investigate the properties of polystyrene, polyvinyl alcohol, and sodium polyacrylate.

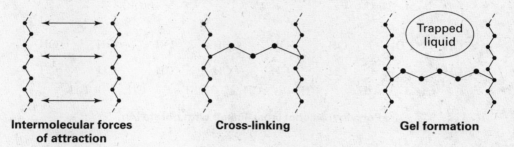

| Intermolecular forces of attraction | Cross-linking | Gel formation |

Polystyrene is not cross-linked, but its intermolecular forces of attraction make it useful for constructing products such as radio cases, toys, and lamps. When polystyrene is expanded to produce the material called plastic foam, it has a very low density and is used to make egg cartons, insulation, and fast-food containers. The intermolecular forces of attraction in polystyrene are destroyed by the action of acetone, and the polymer loses its shape and becomes fluid.

Polystyrene chain

Polymers *continued*

Polyvinyl alcohol can be weakly cross-linked with the hydrated borate ion, $B(OH)_4$. This polymer forms a non-Newtonian gel, which has properties similar to the Slime toy manufactured by Mattel, Inc. When kept in motion, it forms a semi-rigid mass; when held steady, it flows.

Polyvinyl alcohol cross-linked with borate ion

Sodium polyacrylate is a strongly cross-linked polymer that has superabsorbent properties. It can form a gel by absorbing water that has as much as 800 times its own mass. Currently, it is used to coat seeds before planting and to remove water from diesel and aviation fuels. Some brands of disposable diapers contain this superabsorbent polymer.

OBJECTIVES

Infer chemical structures from differences in chemical properties.

Deduce the solubility of plastic foam.

Describe the properties of three polymers.

MATERIALS

- 4% polyvinyl alcohol
- 4% sodium borate solution
- 10 mL graduated cylinder
- 50 mL beaker
- 5 mL acetone
- foam (polystyrene) cup or foam packing peanuts
- glass stirring rod
- medicine dropper
- NaCl
- paper towel
- petri dish
- sodium polyacrylate
- spatula
- watch glass

Polymers *continued*

For this experiment, wear safety goggles, protective gloves, and a lab apron, to protect your eyes, hands, and clothing. If you get a chemical in your eyes, immediately flush the chemical out at the eyewash station while calling to your teacher. Know the locations of the emergency lab shower and the eyewash station and the procedures for using them.

Do not touch any chemicals used in the laboratory. If you get a chemical on your skin or clothing, wash the chemical off at the sink while calling to your teacher. Make sure you carefully read the labels and follow the precautions on all containers of chemicals that you use. If there are no precautions stated on the label, ask your teacher what precautions you should follow. Do not taste any chemicals or items used in the laboratory. Never return leftovers to their original containers; take only small amounts to avoid wasting supplies.

Call your teacher in the event of a spill. Spills should be cleaned up promptly, according to your teacher's directions.

Never put broken glass in a regular waste container. Broken glass should be disposed of properly.

Procedure

1. Measure 5 mL of acetone and pour it into a petri dish. Place two or three pieces of expanded polystyrene (foam) into the acetone. After a few minutes, examine the polystyrene.

Observations:

_____

_____

_____

2. Use a spatula to lift out the polystyrene pieces and place them onto a paper towel. Using a medicine dropper, place a few drops of acetone from the petri dish onto a watch glass and allow them to evaporate. Dispose of the remaining acetone in the metal safety can provided by your teacher. Additionally, place several drops of pure acetone onto a watch glass and allow them to evaporate.

Observations:

_____

_____

_____

109

3. Pour 25 mL of 4% polyvinyl alcohol into a 50 mL beaker. Slowly stir in 3 mL of 4% sodium borate solution. Be sure you have your gloves on. Pour the gel onto your lab bench top. Pick up the gel and knead it into a ball. Pull it slowly, as shown in **Figure A.** Pull it quickly. Hold part of it in your hand over a beaker, and let the rest stretch downward.

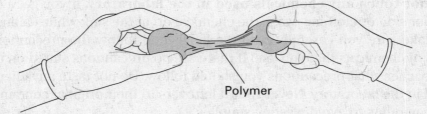

Polymer

Figure A

Observations:

_____

_____

CAUTION Although sodium polyacrylate is nontoxic, it readily absorbs water. For this reason, its dust should not be inhaled.

4. Place two squares of paper towel, about 10 × 10 cm, on your lab bench, about 5 cm apart. Sprinkle the center of one of the paper towels with sodium poly-acrylate polymer, as shown in **Figure B.** Cover each paper towel with a second paper towel.

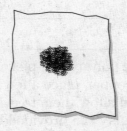

Figure B

5. Slowly add 5 drops of water from a medicine dropper onto the center of the first pair of paper towels, and then slowly add 5 drops of water onto the center of the other pair. Repeat this procedure until one pair of the paper towels appears saturated with water. Remove the top paper towels and examine each bottom towel. Record your observations of the bottom paper towel.

Observations:

_____

_____

Polymers *continued*

6. In the center of a clean petri dish lid, lightly sprinkle a few grains of sodium poly-acrylate polymer. Slowly add a few drops of water until a gel forms. Pause between each drop to see how your gel is growing. By repeating this process, form a tower of gel 2 cm high. Your tower may be more of a mound, depending on how the gel grows. Then sprinkle a few grains of salt onto the top, as shown in **Figure C.**

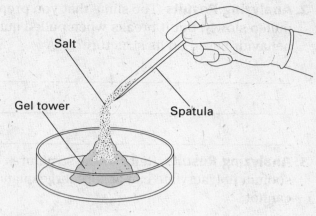

Figure C

Observations:

_____

_____

_____

_____

DISPOSAL

7. Clean all apparatus and your lab station. Return equipment to its proper place. Dispose of chemicals and solutions in the containers designated by your teacher. Do not pour any chemicals down the drain or in the trash unless your teacher directs you to do so. Wash your hands thoroughly after all work is finished and before you leave the lab.

Analysis

1. Analyzing Results From your observations of the residue formed from the evaporated pure acetone and acetone from the petri dish, what can you conclude about the solubility of the foam in acetone?

_____

_____

_____

_____

Polymers *continued*

2. **Analyzing Results** The slime that you prepared in step **3** stretches when pulled slowly, but it breaks when pulled quickly. How can you explain this behavior based on its structure?

_____

_____

_____

3. **Analyzing Results** Refer to the structures of the polymers, and explain why sodium polyacrylate can absorb large quantities of water but polystyrene cannot.

_____

_____

_____

_____

Conclusions

1. **Relating Ideas** What is the advantage of coating a seed with sodium polyacrylate polymer before planting?

_____

_____

_____

2. **Relating Ideas** The polymers you investigated in this lab are only a small fraction of the polymers you encounter every day. Natural polymers are found in your body, in the foods you eat, and in plant structures. Name some polymers that are familiar to you.

_____

_____

Skills Practice

Measuring the Iron Content of Cereals

All foods contain at least one of six basic nutrients essential for organisms to grow and function properly: carbohydrates, proteins, lipids, vitamins, minerals, and water. Few foods contain all six nutrients, and most foods contain a concentration of only one or two of these nutrients. But it is important that the foods we eat contain a combination of all six nutrients, so manufacturers fortify foods with additional vitamins and minerals. This is one way to ensure a properly balanced diet and to prevent nutritional deficiency diseases.

Flour and products made with flour were first fortified with iron in 1940. The human body requires iron for many functions. Most importantly, iron is used in the production of hemoglobin molecules in red blood cells. The iron contained in hemoglobin attracts oxygen molecules, allowing the blood cells to carry the iron throughout the body. Because red blood cells are being replaced constantly, there is a continuous need for iron in the diet. While an iron overdose is harmful and can cause kidney damage, iron deficiency can lead to anemia, a condition in which red blood cells cannot carry a sufficient amount of oxygen to cells. Iron is found in foods such as liver and other red meats, raisins, dried fruits, whole grain cereals, legumes, and oysters.

The nutritional content in packaged foods is listed on the nutrition facts label found on most food packages. The label provides a listing of all the nutrients, along with their recommended dietary allowances (RDAs) based on a 2000 calorie diet. The RDAs are the average daily amounts of various nutrients that, in the opinion of scientists, we should consume daily to remain healthy.

One staple breakfast food is cereal. Most brands provide a minimum of 25% of the RDA for iron. Typically, very tiny particles of pure, powdered metallic iron are mixed in the cereal batter along with other additives. When consumed, the iron particles react with the digestive juices in the stomach and change to a form easily absorbed by the human body. Acidic foods can actually increase iron absorption threefold to sevenfold.

In this investigation, you will isolate and quantify the amount of iron added to three breakfast cereals and compare the amount collected to the amount listed on each food label. In addition, you will simulate how an acidic environment in the stomach aids in the conversion of iron particles to a form easily absorbed in the body.

OBJECTIVES

Isolate and **measure** the iron content in breakfast cereals.

Compare the iron content listed on various cereal boxes to lab results.

Observe the action of acids on iron.

| Measuring the Iron Content of Cereals *continued*

MATERIALS

- 0.2% HCl
- cereals, 3 brands with high iron content, one serving of each
- 400 mL beakers, 3
- balance, graduated in centigrams
- strong bar magnet
- glass stirring rod
- wash bottle, containing deionized water

- magnetic stirrer and stirring bar (optional)
- magnifier (optional)
- mortar and pestle
- test tube
- transparent tape
- wax pencil
- weighing boat

  Always wear safety goggles and a lab apron to protect your eyes and clothing. If you get a chemical in your eyes, immediately flush the chemical out at the eyewash station while calling to your teacher. Know the locations of the emergency lab shower and the eyewash station and the procedures for using them.

 Do not touch any chemicals. If you get a chemical on your skin or clothing, wash the chemical off at the sink while calling to your teacher. Make sure you carefully read the labels and follow the precautions on all containers of chemicals that you use. If there are no precautions stated on the label, ask your teacher what precautions you should follow. Do not taste any chemicals or items used in the laboratory. Never return leftovers to their original containers; take only small amounts to avoid wasting supplies.

 Call your teacher in the event of a spill. Spills should be cleaned up promptly, according to your teacher's directions.

 Never put broken glass in a regular waste container. Broken glass should be disposed of properly.

Procedure
PART 1: ISOLATING IRON FROM BREAKFAST CEREALS

1. Record in the **Data Table** the name of each of three iron-fortified cereals. Read the nutrition facts label for each cereal to find the mass of one serving and the percentage of the RDA for iron that is in each serving. Record these measurements in the **Data Table.**

2. Use a wax pencil to label three beakers, each with the brand name of one of the cereals. Weigh one serving of each cereal and put each into its labeled beaker.

3. Use a mortar and pestle to grind a serving quantity of each cereal into a fine powder. Place each powdered cereal in the appropriately labeled beaker. Remember to *clean* the mortar and pestle *after each use.*

4. Determine the mass of a bar magnet to the nearest milligram, and record your results in the **Data Table.**

5. Tape a bar magnet to the end of a glass stirring rod, leaving most of the bar exposed.

6. Pour 300 mL of warm water into one of the three beakers. Continuously stir the powdered cereal with the bar magnet for 10–15 min. The longer the cereal solution is stirred, the more the tiny iron particles will precipitate out of it and attach to the magnet. About 30 min of stirring gives the best results.

7. Remove the magnet, gently rinse it with deionized water, and carefully drain any excess liquid from it. When the magnet is dry, determine the mass of the magnet and iron slivers to the nearest milligram. Record your results in the **Data Table** and on the chalkboard so that average class results can be determined.

8. Place your bar magnet on a paper towel. Use a magnifying lens to observe the tiny slivers of iron attached to the bottom of the magnet. Record your description of the iron slivers below. Then collect the iron slivers from the magnet and save them for step **9.** Repeat steps **4–7** for the remaining two cereals.

Observations:

_____

_____

_____

_____

PART 2: SIMULATING THE ABSORPTION OF IRON IN THE BODY

9. Combine in a test tube all of the iron slivers you collected.

10. Pour 5 mL of 0.2% HCl in the test tube, and let the iron slivers and HCl stand for 10 min. Record any change(s) in the color, shape, or size of the iron slivers.

_____

_____

_____

_____

DISPOSAL

Clean all apparatus and your lab station. Return equipment to its proper place. Dispose of chemicals and solutions in the containers designated by your teacher. Do not pour any chemicals down the drain or put them in the trash unless your teacher directs you to do so. Wash your hands thoroughly after all work is finished and before you leave the lab.

Measuring the Iron Content of Cereals *continued*

Data Table

	Cereal A:		Cereal B:		Cereal C:	
	Your result	Class average	Your result	Class average	Your result	Class average
Initial mass of magnet (g)						
Final mass of magnet (g)						
Mass of iron recovered (mg)						
Percentage of RDA for iron on the label						
Accepted mass of iron (mg)						
Percentage error						

Analysis

1. Organizing Data Calculate the mass of iron recovered from each cereal by subtracting the initial mass of the magnet from the final mass of the magnet. Record your answers in the **Data Table.**

_____

_____

2. Organizing Data Calculate the accepted mass of iron in each cereal tested by multiplying the percentage from the nutrition facts label by 18 mg (the RDA for iron). Record your answers in the **Data Table.**

_____

_____

3. Organizing Data To determine the percentage error for each iron measurement, subtract the mass of iron recovered from the accepted mass of iron. Then divide the difference by the accepted mass of iron, and multiply by 100%. Record your answers in the **Data Table.**

Measuring the Iron Content of Cereals *continued*

4. Applying Concepts A food label lists the serving amount of a particular food as providing 35% of the RDA of iron. If the RDA based on a 2000 Cal diet is 18 mg, what amount of iron must a person consume from other foods to get 100% of the recommended dietary allowance?

_____

_____

_____

5. Analyzing Data Examine your percentage error entries in the **Data Table.** How closely do your results verify the claims made by the cereal manufacturers? If your percentage errors are high, suggest reasons for this.

_____

_____

_____

5. Inferring Conclusions How do you think the iron isolated from cereal compares with the iron found in nails and bridges?

_____

_____

_____

Conclusions

1. Designing Experiments Use a metal file to scrape off metal flakes from a piece of iron metal or nail. Compare the size of the particles collected from this iron to the particle size collected from cereals.

_____

_____

_____